Light Power:
Half a Century of Solar Electricity Research

Volume 1: Solar Thermal Power Systems

Light Power:
Half a Century of Solar Electricity Research

Volume 1: Solar Thermal Power Systems

editor

David Faiman

Ben-Gurion University of the Negev, Israel

World Scientific

EW JERSEY · LONDON · SINGAPORE · BEIJING · SHANGHAI · HONG KONG · TAIPEI · CHENNAI · TOKYO

Published by

World Scientific Publishing Europe Ltd.

57 Shelton Street, Covent Garden, London WC2H 9HE

Head office: 5 Toh Tuck Link, Singapore 596224

USA office: 27 Warren Street, Suite 401-402, Hackensack, NJ 07601

British Library Cataloguing-in-Publication Data
A catalogue record for this book is available from the British Library.

ISBN 978-1-78634-756-5

For any available supplementary material, please visit
https://www.worldscientific.com/worldscibooks/10.1142/Q0224#t=suppl

Desk Editor: Cheryl Heng

Typeset by Stallion Press
Email: enquiries@stallionpress.com

Preface

History is usually written by people who did not themselves witness the events they document. And on those occasions in recent times when it has been possible to witness events of historic value it is often non-specialists, such as journalists, who perform the documentation. This latter fact has its virtues in that a more balanced picture may result compared to what might otherwise have been the case had the history been written by a single participant. However, the downside of a non-specialist history is frequently a lack of important details.

The advent of large scale solar power generation is certainly a historic event that has come about within living memory and one which continues to develop as these words are being written. Furthermore, the author, although not himself having contributed anything of historic value to the subject, has been intimately involved in the research and development of solar power, and been fortunate to have associated with many of the scientists and industrialists who have brought about it. In particular, for some thirty years he hosted a sesquiennial symposium on solar power generation, which served as a platform for many of the most influential people whose creations did make significant contributions to the present situation.

In those symposia, balance was provided by the inclusion of all solar power technologies within a single meeting which would last two or three days. And there were no parallel sessions. Thus, a unique feature of the *Sede Boqer Symposia* (referring to the venue at which they were held) was that all participants, no matter what their particular specialty, were able to attend all presentations. This fact placed an important constraint on the keynote speakers: Namely, that the first 15-minutes of their otherwise specialist presentation had to be in a language that all could understand. "All" in this situation referred to scientists of other disciplines, but equally important, to industrialists and government decision makers. Those keynote lectures, which were originally published as booklets by Ben-Gurion University of the Negev, and distributed, some months later, to all the attendees, thus constitute a record of progress, in the actual words of the people who were responsible for much of it, and, equally important, in a format that is accessible to a wide range of people.

This was the motivation that led the present editor to republish a selection of the keynote lectures in the form of a history in the words of those who made it. The original idea was to reproduce those lectures in strict chronological order, maintaining the year-by-year mix of photovoltaic and solar-thermal technologies, in order to emphasize the manner in which interest in each has ebbed and flowed throughout the past half century. However, it turned out that to do so would have rendered this book to be unreasonably lengthy. So, instead, a separation has been made between the solar thermal presentations, which constitute the present "Volume 1" of this history, and the photovoltaic (PV) presentations which will follow later. It makes sense to create such a separation, because large-scale solar-thermal power production got off to an earlier start than did PV, and is now relatively mature. On the other hand, PV, which started with flat panels of high-cost silicon cells of low efficiency, has since branched out into what appears to be a never-ending exploration of new materials: and one, moreover that continues to the present day and will probably not stop here.

Thus, to summarize: This is a specialist's history of solar-thermal power generation, edited by a specialist, but with balance maintained by including the opinions of all the competing specialists whose creations led to the present commercialization of the field. Nevertheless, interested non-specialists should be able to read it by virtue of the special constraint, mentioned above, that was placed upon the speakers whose lectures constitute the bulk of this book.

Acknowledgments

The editor acknowledges Ben-Gurion University of the Negev, the copyright-holders of the symposium proceedings volumes, from which most of the material in this book has been extracted. He is deeply indebted to Ms. Roxana Dann, for scanning the pre-digital presentations from the older among those volumes, converting the scans to editable files, and performing a first proof-reading of the entire contents of this book. He thanks: Benny Doron and Micky Margalith for correcting his translations of their original Hebrew-language papers; and Manuel Collares Pereira, Jan Kreider, Roland Winston, Wilfried Grasse, Paul Klimas, Tom Mancini, David Mills, Aldo Steinfeld, and Chemi Sugarmen for their agreement to have their historic presentations reproduced in this volume. The historic presentations of Ari Rabl and Harry Tabor are reproduced without permission as their authors are sadly no longer with us. May these contributions act as tributes to their memories. In the case of Dr. Rabl's presentation, the editor acknowledges the help of Mr. Xu Ke, of World Scientific Publications, for converting a sound recording into text files. More generally, the editor wishes to thank the staff of World Scientific Publishers for the excellent work they have performed on restoring several of the old figures that are reproduced in this book. Finally, every effort has been made to identify and acknowledge the copyright holders of the many figures that are reproduced within the presentations of the various contributors to this book. Where errors have occurred, the editor offers his profound apologies and will include any requested errata in volume 2.

Contents

Introduction

In recent years, *energy* has become an increasingly mouthed word in many contexts. The widespread awareness of its economic importance only commenced towards the end of the twentieth century. Specifically, in 1973 there occurred a so-called "oil crisis" in which people at large first began to become conscious of the price of oil and its effect on both the economy and the environment. Prior to that time the cost of oil had been so low that it had largely replaced coal as a heating fuel and for electricity generation. Coal was inexpensive, but oil was very much cleaner, resulting in far less pollution when it was burned. Oil was also more convenient and cheaper to transport than coal. As a result of this change-over in fuel preference, those infamous fogs that had characterized London and many other cities up through the 1950s soon became a thing of the past. Furthermore, the gasoline derivative of oil could be manufactured at such low cost that, in those days, nobody was seriously concerned with the efficiency of automobile engines.

Such a lackadaisical attitude to energy changed abruptly in 1973 when, as a result of various political forces on the world stage, the price of a barrel of crude oil jumped in a matter of weeks, and rose by a factor of ten (in today's dollars) in a mere ten years. This was the "crisis" alluded to above. Figure 1 shows the price of oil on a year-by-year basis in US$ (adjusted to its 2015 dollar value) [1].

The enormous price increase in such a basic commodity as oil naturally had economic repercussions throughout the energy industry and everything connected to it — which means essentially everything. Electricity generation became more expensive, as did the cost of

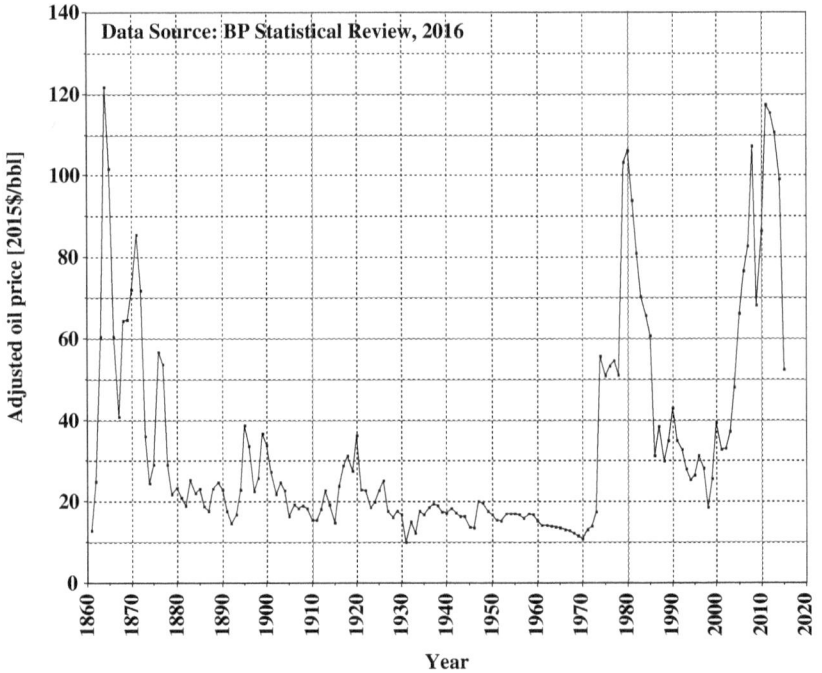

Figure 1:   History of oil prices from its discovery until the present, adjusted to the 2015 value of the US$ [1]

heating fuel and transportation. All of a sudden, people became conscious of the need for energy economy: smaller automobiles became popular, with engines that burned fuel more economically; insulation was added to buildings, to cut down on the amount of winter heating and summer cooling that would be needed. Above all, governments started to take an interest in the development of renewable energy resources such as solar and wind power.

This sudden interest in energy saving and renewable resources was, of course, fueled by economic interests. However, something in parallel was also happening to the Earth's atmosphere. The concentration of a trace gas, carbon dioxide, which at the beginning of 1970 had stood at 325 parts per million (ppm), was silently rising. In 2015 it crossed the 400 ppm level (Fig. 2): an increase of nearly 25% in less than half a century, and it is still rising [2].

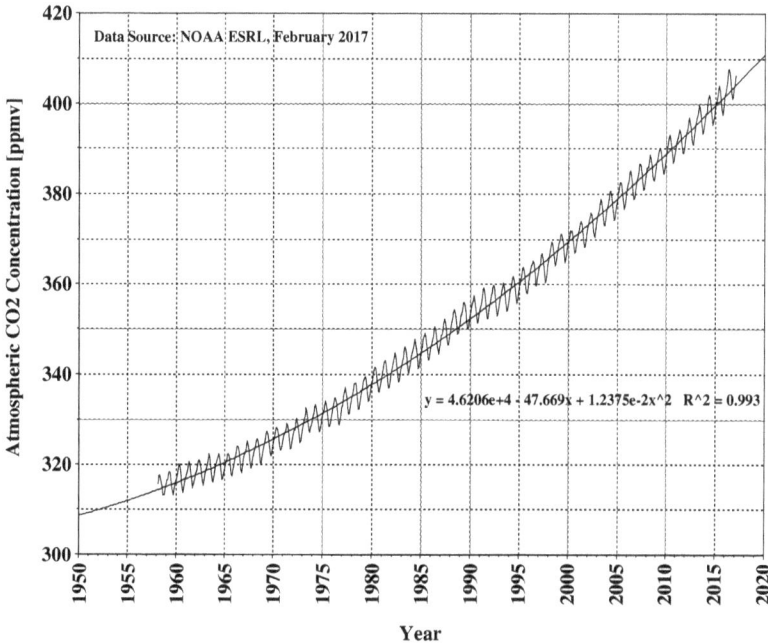

Figure 2: Data from atmospheric carbon dioxide concentration measurements on Mt. Mauna Loa [2], fitted by the present editor to a smooth quadratic curve

Now, such a small concentration of gas in the atmosphere (four hundredths of 1%) might not sound like anything that need concern us. Unfortunately, this is not the case. The reason is that the delicate balance between how much radiation the Earth absorbs from the Sun and how much it emits back into outer space is controlled by the trace amounts of water vapor and carbon dioxide in the atmosphere — the so-called *greenhouse effect*. This balance, depending on which way the trace amount shifts, can tend towards a completely frozen Earth in which no liquid water could exist or, in the other direction, a planet that would be too hot to support life. In either case, our planet would be unfit for any conventional life forms. Hence, there is certainly reason to be concerned that in a mere half-century, the atmospheric carbon dioxide concentration, small as it is relative to the principal gases in our atmosphere, has increased by 25% and shows no sign of diminishing.

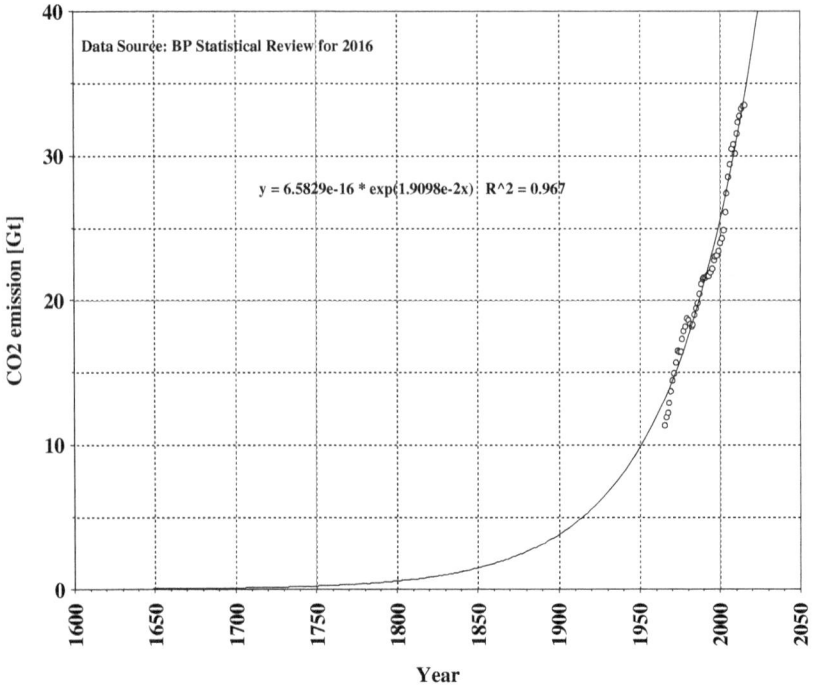

Figure 3: Data of annual carbon dioxide emissions since 1970 [1] with the editor's exponential least-squares fit

Before people started to use fuel for purposes other than cooking and keeping warm, the natural carbon dioxide concentration is estimated to have been approximately 300 ppm. However, in the middle of the 18th century, the so-called *Industrial Revolution* resulted in the emission of increasing quantities of carbon dioxide due to the burning of ever greater quantities of fuel for transportation and various industrial purposes. Moreover, from detailed accounting that has been made since the middle of the 20th century, the rate of increase in carbon dioxide emission appears to be following an exponentially rising curve (Fig. 3).

At this stage of the discussion, it is important not to become blinded by what may simply be some numerical coincidences rather than cause-and-effect connections. First, it may simply be a coincidence that an exponential fit to relatively recent annual emission data [1] extrapolates back in time to a level compatible with the start of the industrial revolution. A quadratic curve would fit the actual data equally well. However, instead of diminishing rapidly to zero as we look back in time, such a curve would turn around near the year 1935, rising again and reaching present-day emission levels back in 1855. Such a quadratic curve most certainly does not represent the emission of fossil fuel before the mid-twentieth century. What is certain, however, is that carbon dioxide emissions were small by modern standards before the industrial revolution and have been rising since then. So an exponential fit to the data is probably a reasonable description — it is certainly better than a quadratic fit — but one must still beware of assuming that it will continue into the future. *That much, at least, may still be in our hands.*

A second possible coincidence is the connection between rising fossil fuel emissions and the rise in atmospheric carbon dioxide. It may be the case — for nobody knows for certain — that some extra-terrestrial phenomenon is causing the planet to warm up, and that the corresponding increasing temperature of the oceans is causing the latter to release larger quantities of carbon dioxide into the atmosphere. After all, the oceans contain vast quantities of dissolved carbon dioxide, and the solubility of this gas decreases with rising water temperature. Such a situation would clearly not be in our hands to alter. However, the ever increasing burning of fossil fuel would still exacerbate the situation. Hence, whatever one's stand on the so-called "global warming" issue may be, it should certainly be important to reduce fossil fuel emissions as much as possible, both of carbon dioxide, and of the accompanying polluting emissions that have a more immediate effect on health and the environment.

Thus far, our general introduction has summarized the energy issue as it came to be widely accepted after the Earth Summit conference that was held at Rio de Janeiro in 1992. This is in order to provide a sense of perspective for the various presentations that form the bulk of the rest of this book. Most of the latter were given at a time when the emphasis of solar power research was on trying to enable it to replace high-cost fossil fuel, with less emphasis on its environmental importance.

So, what was the situation before Rio? Back in the 1980s, the largest solar-thermal plant already had a power rating of order 10 megawatts but the associated parabolic trough technology turned out to be sufficiently robust that it actually gave credibility to a future that could be powered by solar energy. As a result, solar-thermal plants were quite quickly deemed feasible on the 100 megawatt scale, employing parabolic trough and a number of other emergent technologies. To a certain extent, further scientific research and development became almost irrelevant as a host of commercial manufacturers competed with one another for shares in an increasing market.

By comparison, large-scale photovoltaic power production, which has also moved from being a dream to an on-going reality, had a more modest start. Back in 1986, the largest photovoltaic solar power plant had a 5 megawatt rating, but its performance degraded rapidly, causing the plant to be shut down and dismantled. However, system reliability improved to such a degree that by the second decade of the 21st century, photovoltaic plants were being financed with guaranteed performance lives of 25 years or more, and power ratings approaching the gigawatt scale.

One might consequently ask: "With solar power generating technology so firmly established on a commercial basis, why do we look back at these 'ancient' presentations?"

There may, of course, be a certain fascination at reading the words of these specialists fresh "from the horse's mouth" as it were. However, were the reason to be merely one of historical interest, it would not be necessary to reproduce these papers in all of their technical detail. There is, however, a second reason for delving into history in this manner. Namely, most of the present authors were major actors on the stage of solar development. Some of the technologies they discussed became a reality while others did not. For the most part, those technologies that did not succeed fell foul of economic or technical issues that were relevant at the time but which may have changed in the meantime, or will do so in the future. Therefore, instead of having to re-invent the wheel, it may be of value to re-read these presentations as they were first given by the people who designed and fabricated those components and systems. Some of the issues that were raised in the question and answer sessions back then, seen as insurmountable barriers at the time, may no longer be so.

There is also a third point that is important to emphasize when looking back at the various solar technologies that are covered in these presentations. Namely, when treated purely as the economics of one technology versus that of another — as is natural for manu-facturers who need to sell a product — it is easy for us to lose sight of the fact that power technologies of differing costs *have to* co-exist. This is because utilities must generate round-the-clock electricity and do so *economically*, even if no renewable technologies are involved. In order to do so they employ a mix of various technologies, each having different reaction times, fuel types, and maintenance costs, but operated in a manner that closely monitors varying customer demands. For example, in principle, gas turbine generators could meet all of the technical requirements for efficient 24 hours per day generation, but they would not be economical for continuous use. Similarly, with increasing input from renewable technologies, the different time responses of, say, photovoltaic and solar-thermal generators, could cause the integrated economic value of the two to far exceed their relative dollar-per-watt operation cost. This point is relevant not only to the relatively limited number of solar technology types that are presently under large-scale construction, but may possibly also be pertinent to some of the once abandoned solar thermal technologies discussed in the present volume, and the photovoltaic ones in the follow-up volumes.

In this connection, it will be noticed that the emphasis in the early presentations reproduced here, and in the corresponding discussions, is almost entirely on simple economic questions such as how long it takes for a specific solar energy system to pay off its initial investment via fuel savings. It is only with hindsight that we can now add total system considerations to what these pioneers have bequeathed us, together with the realization that system-versus-system economics as it was understood back then, needs to be re-examined in the broader context of their respective time-integrated contributions to future electricity grids.

References

1. BP Statistical Review, 2017.
2. NOAA, ESRL, February 2017.

Chapter 1

Solar Power from the Perspective of 1986

1.1 Editor's Foreword

The series of lectures upon which this book is based, commenced in 1986. However, the development of potential solar power-generating technologies had started much earlier. Two popular books have documented the early history of solar-thermal power [1], and the development of photovoltaic cells for use in spacecraft before their eventual deployment on Earth [2]. Thus, when the oil crisis of 1973 occurred there was already a technological base upon which to build. This chapter accordingly starts with a review of all the pre-1986 solar power plants, which the present editor gave at the 1986 symposium as a basis for discussion at that meeting. By the time the series of Sede Boqer Symposia on Solar Electricity Production commenced, several small and not-so-small solar power-producing plants were already in existence. Two, which were of particular interest at that time, were the 5 Megawatt "Solar Pond" plant that Ormat Corporation built at Beit Ha'Aravah in Israel and the 10 Megawatt parabolic trough "SEGS-1" plant that the Luz Corporation built at Dagget in California. At our first symposium, there were no keynote speakers. However, we were fortunate to have short presentations on the first year's operating experience and performance of both of these plants. Those presentations were given in Hebrew, but English translations, that were prepared by the present editor, and approved by the original authors, are given in this chapter.

References

[1] Butti, Ken and John Perlin 1980. *A Golden Thread: 2500 Years of Solar Architecture and Technology* (Cheshire Books, Soquel, CA).

[2] Perlin, John. 1999. *From Space to Earth: The Story of Solar Electricity* (aatec Publications, Ann Arbor, MI).

1.2   There was Solar Power before 1986 – A Review (Prof. David Faiman)

A presentation by Professor David Faiman (Jacob Blaustein Institutes for Desert Research, Ben-Gurion University of the Negev, Sede Boqer Campus, Israel) at the First Sede Boqer Symposium on Solar Electricity Production, 23–24 February, 1986.

1.2.1   *Introduction*

The following pages present a review of the solar electricity "power stations" that have been built in various parts of the world. There are about forty in number if we restrict our discussion to stations of 100 kWp or greater (the size of the installations which the Israel Ministry of Energy and Infrastructure would like to investigate at the National Solar Electricity Technology Test Facility at Sede Boqer). The review starts with a few basic concepts concerning the conversion of solar energy into electricity. It continues with a discussion of the power plants that have been built in various parts of the world, this part being subdivided into sections on solar thermal facilities of the saline pond, central receiver, concentrating dish and concentrating trough varieties, and photovoltaic facilities. The review concludes with some observations regarding potential land requirements together with some general recommendations as to what might usefully be done at Sede Boqer and other new solar test centers.

1.2.2   *Solar resource availability*

Several of the facilities reviewed below turned out to produce substantially less electricity than expected because the insolation statistics of the site were not well known. Fortunately, thanks to the efforts of the Blaustein Institute's Desert Meteorology Unit over the past several years, this will not be a problem at Sede Boqer. A useful starting point, therefore, would be to examine the amounts of insolation that would be available at Sede Boqer for the various kinds of technology. The relative quantities that we shall discuss should also be approximately relevant to other desert test sites.

Our interest here is in the total amount of energy that would be available in the course of a year, depending upon the tracking mode of the collector and its concentration ratio. Table 1.2.1 lists the various amounts on a relative scale where "1.00" represents about $2000\,\mathrm{kWh/m^2/year}$. A few examples from the table will suffice to illustrate its use.

The figure "1.00" corresponds to the annual global insolation on a horizontal plane. It represents the amount of energy "seen", for example, by a solar pond. This comprises 100% of the annual diffuse radiation together with an integral of all appropriate projections of the beam component, the latter, however, arriving for the most part at an unfavorable incident angle.

The situation may be improved by having the collector track the sun's apparent position in the sky. The best possible collector would be a non-concentrator (so as not to lose the diffuse component) which

Table 1.2.1:   Annual solar availability for various kinds of collector at Sede Boqer

Configuration	Index	Examples
Tracking concentrator (E-W axis)	0.79	Parabolic troughs
Tracking concentrator (N-S axis)	0.96	Parabolic troughs
Horizontal flat plate	1.00	Solar ponds
Tracking concentrator (polar axis)	1.03	"short" troughs
Tilted flat plate (tilt = latitude)	1.07	PV cell arrays
Tracking concentrator (two axis)	1.08	Parabolic dishes, Fresnel lenses
Tracking flat plate (E-W axis)	1.12	PV cell arrays
Tracking flat plate (N-S axis)	1.26	PV cell arrays
Tracking flat plate (polar axis)	1.33	PV cell arrays
Tracking flat plate (two axis)	1.37	PV cell arrays

(based upon 1982/83/84) beam and global data with
isotropic diffuse assumption)

follows the sun both in azimuth and in altitude. From the table, 37% more energy would be available on an annual basis.

Matters are almost as good if single-axis "polar" tracking is employed. Here a N-S tracking axis is tilted towards the equator at an angle to the horizontal equal to the latitude of the site (30.85° N for Sede Boqer). Polar tracking is however not suitable if the collector is very long. In such circumstances a horizontal axis must be employed, and on an all-year basis, this axis should point north-south.

Lastly, the table indicates the heavy energy penalty (even at Sede Boqer) that results from rejecting the diffuse component, as one must when a concentrating collector is employed. For example, a north-south axis tracking flat plate collector (e.g. a photovoltaic panel) would collect some 30% more energy than an equal aperture parabolic trough concentrator employing the same tracking strategy.

1.2.3   *Low temperature solar thermal facilities*

A number of intrinsically low temperature solar technologies are under study in Israel. These include: saturated solar ponds (Tushia Corp), Fresh water solar ponds (Arel-Argaman Corp), and the more-developed saline solar ponds (Ormat Corp). Only the latter type have been used, to date, for electricity production.

1.2.3.1   *Saline solar ponds*

Solar ponds are intrinsically low-efficiency devices. Because the saline variety are "unglazed" their thermal efficiency is limited to 20% or thereabouts. Moreover, on account of the low temperatures involved, the subsequent conversion of thermal to electrical energy is down at the 5% level. (The Carnot efficiency for a 97°C source and a 27°C sink is 19% but more realistic analyses [1] of the Rankine cycle employed by such turbines indicate much lower efficiencies). Accordingly, the total solar-to-electricity efficiency of such power stations will be of the order of 1% or so.

However low the efficiency of such technology, its great advantage over all competitors lies in the fact that it has built-in storage. For the example shown in Fig. 1.2.1, a hypothetical pond of area

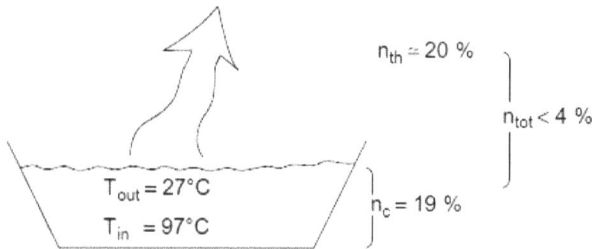

$$n_{th} \approx 20\ \%$$

$$n_{tot} < 4\ \%$$

$$n_c = 19\ \%$$

$$T_{out} = 27\,°C$$
$$T_{in} = 97\,°C$$

Example: $l_{inc} = 2000\ kWh/m^2/year$

       $n_{tot} = 1\ \%$

       $A = 1\ km^2$

       $\bar{P} = 2.28\ MW$ (continuous)

        or 13.7 MW (4 hours/day)

Figure 1.2.1:   A hypothetical solar pond of area 1 sq.km

Table 1.2.2:   Solar pond power stations

System	Aperture [m²]	Peak Power [KW]
Ein Bokek	7,500	150
Beit Ha'Aravah	250,000	5,000 (planned)

1 sq.km and efficiency 1% would produce, on average, only 2.28 MW of electricity, but this output would be truly continuous — day and night, winter and summer. Furthermore, from experience gained by Ormat Corp at their Ein Bokek pond, it is possible, during periods of peak requirements to draw substantially larger amounts of electricity than the average power of the pond. For example, the 2.28 MW pond of Fig. 1.2.1 could produce 13.7 MW for periods of 4 hours per day.

Table 1.2.2 lists the only two solar pond power stations that have been built to date: both being in Israel. Performance details will be found in the contribution to these proceedings by Eng. Benny Doron.

1.2.4 *High temperature solar thermal facilities*

To achieve higher Carnot efficiencies it is necessary to concentrate the incoming solar radiation in order to produce higher temperatures. Two factors limit the temperature that can be reached: the effective temperature of the sun itself, about 6000 K, and the diameter of the solar disc. This latter limits the concentration ratio of line focusing collectors (e.g. troughs) to 213×, and of point focusing devices (e.g. dishes) to 45,300× [2].

However, there is a competing factor. Higher temperatures produce higher Carnot efficiencies but larger heat losses. It is thus necessary to optimize the range of operating temperatures for any given kind of solar collector so as to maximize the total solar-to-electricity efficiency. Figure 1.2.2 indicates some calculated [3] optimum operating temperatures and the resulting total efficiencies to be expected from power stations of the central-receiver variety.

1.2.4.1 *Central-receiver power stations*

Six central-receiver electricity-generating facilities have been built to date. They are listed in Table 1.2.3. Each has been extensively documented in the scientific literature, therefore no attempt will be made to discuss these projects in any depth here. Instead, I shall emphasize some specific lessons that should be of relevance to our program at Sede Boqer and any new programs elsewhere. These comments, which are mainly negative, should in no way be construed as an attempt to pass judgement on scientific programs that involved untold numbers of man-hours of design, execution, operation and careful documentation. Neither should they be interpreted as an indication of personal preference for one kind of technology over another. The inclusion of such, comments serves merely to draw attention to design weaknesses that were reported by the original pioneers in order that others coming later — for example, ourselves — should not repeat them.

1.2.4.1.1 "CRS-1", Tabernas nr. Almeria, Spain

*** Collectors: 93 Martin Marietta 40 sq.m heliostats.

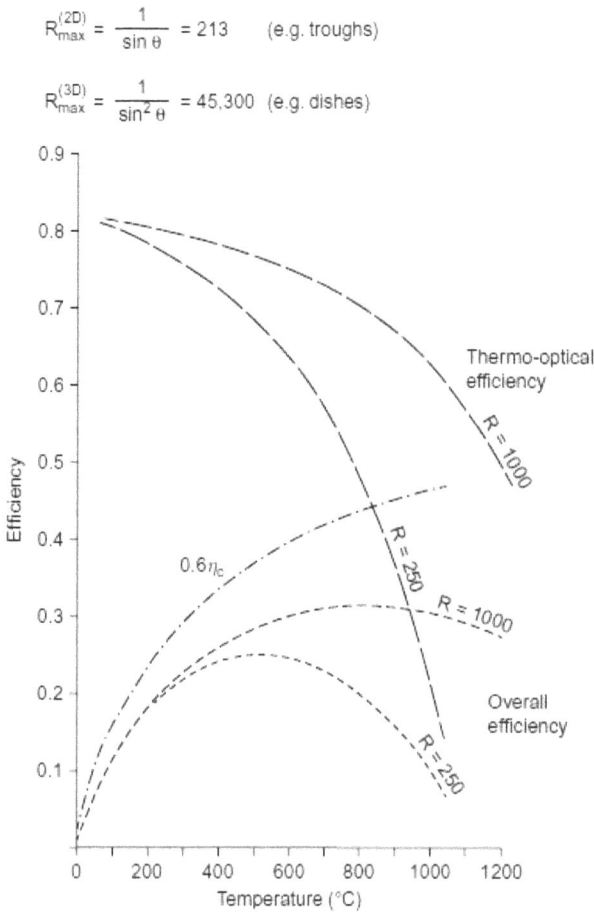

$$R_{max}^{(2D)} = \frac{1}{\sin \theta} = 213 \quad \text{(e.g. troughs)}$$

$$R_{max}^{(3D)} = \frac{1}{\sin^2 \theta} = 45,300 \quad \text{(e.g. dishes)}$$

Figure 1.2.2: Efficiency vs. temperature for two typical levels of solar concentration

*** Specific problems:

— Collector area under-designed as a result of inaccurate insolation data.
— Severe problems caused by sodium leakage necessitated complete redesign of storage tanks.
— Large thermal inertia in steam system: no electricity before noon.
— Large fraction of outages (84%) caused by off-the-shelf (non-solar) components.

Table 1.2.3:   Central-receiver power stations

	CRS-1 Spain	Eurelios Italy	NIO-1 Japan	CESA-1 Spain	Themis France	Solar-1 USA
Peak power [MWe]	0.5	0.6	1	1	2.3	12.1
Field shape	fan	fan	circular	fan	fan	circular
Tower height [m]	43	55	60	60	100	77
Receiver fluid	sodium	water	water	water	hitec	water
Fluid temp [°C]	530	510	249	520	430	510
Fluid press [bar]	2.7	64	39	100	1–2	102
Heliostat area [m^2]	3,655	6,216	12,912	11,400	10,740	72,538
Receiver area [m^2]	9.7	15.9	56.6	11.6	16.0	302.0
Concentration ratio	377	398	228	983	671	240
Global efficiency [%]	16.5	6.6	8.5	12.5	21	15.3
Operational	1981–1984	1981–1984	1981–1984	1983–	1983–	1982–
Project cost [$/W]	30	12	26	15	12	14

— Heliostat reflectance loss of up to 2% per day in absence of frequent (every 3 to 4 weeks) washing.

*** Sources: [4–6].

1.2.4.1.2 "Eurelios", Adrano (Sicily), Italy

*** Collectors: 112 Messerschmitt-Boelkow-Blohm 24 sq.m heliostats 70 Cethel 52 sq.m heliostats.

*** Specific problems:

— Collector area under-designed as a result of inaccurate insolation data. (Design power of 1 MWe reduced to 700 kWe owing to reduced number of heliostats installed, but barely 600 kWe achieved).
— Low heliostat availability (less than 90%).
— Poor mirror reflectance (about 72%). Rapid reflectance loss after cleaning (4–5% after one week, 10% after 3–4 weeks).
— Receiver and thermal cycle failures.
— Excessive storage losses.
— Long start-up phase in the morning.

*** Sources: [7–10].

1.2.4.1.3 "CRT", Nio (Shikoku), Japan

*** Collectors: 807 16 sq.m Japanese heliostats.

*** Specific problems:

— Mirror reflectance decreases by 4% in 2 weeks after cleaning. Six workers take 15 days to hand clean all mirrors. Condensation of moisture on mirror surfaces at night considered the major problem.
— Initially large heat losses from storage owing to poorly insulated support structure.
— Data acquisition system does not provide enough information.
— Parasitic losses judged relatively high. Initially, 17–18% but reduced by 20% as a result of altering pump schedule.

*** Sources: [11, 12].

1.2.4.1.4   "CESA-1": Tabernas nr. Almeria, Spain

*** Collectors: 150 CASA-II 39.6 sq.m heliostats.

             150 SENER 39.6 sq.m heliostats.

*** Specific problems:

— High thermal losses from initial receiver necessitated re-design.

— Miscellaneous technical problems caused delay in experiments.

— Experimental programme presently in progress.

*** Sources: [3, 13].

1.2.4.1.5   "Themis", Targasonne nr. Odeillo, France

*** Collectors: 200 Cethel 54 sq.m primary heliostat.

             11 auxiliary 70 sq.m parabolic dishes

*** Specific problems:

— Serious wind damage to heliostat field caused by unexpectedly high gusts (estimated to have been in excess of 160 km/h).

— Electrical damage caused by thunderstorms in spite of precautions.

— Regulation of the intense solar flux at the receiver is assessed as being "not easy, ... even dangerous".

— Major cause of problems: less than top quality equipment (solar and non-solar) and not enough care taken in design and assembly.

— Experimental programme presently in progress.

*** Sources: [14, 15].

1.2.4.1.6   "Solar-1", Barstow CA, USA

*** Collectors: 1818 Martin Marietta 39.9 sq.m heliostats.

*** Specific problems:

— Lightning storm caused communication and control failures. Necessitated additional grounding protection in field cables.

— Corrosion problems with heliostat assemblies.

— Leak in oil storage tank found to be due to flaw in steel plate material (rather than in a weld).

— Unusually cold weather conditions caused malfunctions, *inter alia*, among monitoring equipment.

— Diurnal temperature cycling caused leaks and other malfunctions.

— Water purification complications on account of severe sensitivity of receiver to water chemistry.

— In the absence of heliostat cleaning, reflectance dropped from 91% to 72% in three months. Thereafter, cleaning (1 min per heliostat using pressurizes demineralised water) restored reflectance to 86%.

— Experimental programme currently in progress.

*** Sources: [16, 17].

1.2.4.1.7 The economics of central-receiver electricity

Regarding the economics of central-receiver electricity, three studies are of particular importance. The first two concern the cost figures of the two existing facilities, "Solar-1" in the USA [17] and "Themis" in France [14]. The third study [18] is a detailed economic projection of lessons learned from "Solar-1" to a planned "Solar-100" project, originally destined for completion in 1990 but subsequently deferred for an indefinite period [19].

Tables 1.2.4 and 1.2.5, taken from Ref. [17], give a detailed cost break-down of "Solar-1" as regards its construction costs and annual operation and maintenance. One sees that this first-of-a-kind project

Table 1.2.4:   Solar-1 capital cost***

Solar facility	Cost [M$]	Percent [%]
Solar Facility Design Cost	31.2	22
Collector Field Fabrication & Construction	40.0	28
Receiver Fabrication & Construction	23.4	17
Thermal Storage Fabrication & Construction	12.0	8
Plant Control System	3.0	2
Beam Characterization System	1.0	1
Miscellaneous Support Systems	9.4	7
TOTAL SOLAR FACILITY		
Design/Fabrication/Construction Cost	120.0	85
Turbine/generator design & construction cost	21.5	15
Total plant cost	141.5	100

Table 1.2.5:   Solar-1 FY83 operating and maintenance budget***

Item	Cost [$]	Percent [%]
Company Labor	2,090,000	62
Material	471,800	14
Contract	288,800	9
Other Miscellaneous Expenses	49,900	2
Administration and General Overheads	469,200	13
Total Annual Cost	3,369,700	100

*** *Source*: Bartel and Skvarna 1983 (SAND83-8021)

cost approximately $14/Wp to execute and that its operating costs amount to some $3.7 M per year. For an estimated annual electricity production of 13 million kWh, this O&M figure would contribute 28c/kWh to the cost of the electricity.

The economics of "Themis" are comparable [14]. The total construction cost was $14.8/Wp and annual O&M amounts to $1 million for a projected annual electricity production of 3 million kWh.

The projected picture for "Solar-100", should that project ever be built, is considerably more favorable [18]. Construction costs are estimated at $5.8/Wp and annual O&M would only contribute about 1c/kWh to the cost of electricity. However, as we shall see below, both parabolic dish and parabolic trough solar thermal systems have already been built for under $5/Wp and this is probably one of the considerations that has led to deferment of "Solar-100".

1.2.4.2   *Distributed-field parabolic dish power stations*

In contrast to the central-receiver concept where the combined outputs of a field of heliostats are all focused on a single receiver, parabolic dish facilities employ separate receivers for each dish or group of dishes. The potential advantages of such collectors for high efficiency electricity generation have been discussed elsewhere [20]. Here, we only point out that three such facilities have been built at the scale of 100 kWe and these are listed in Table 1.2.6. Once again, without intending to pass judgement, our emphasis will be on system problems.

Table 1.2.6: Parabolic dish power stations

	Sulaibiya Kuwait	Shenandoah USA	Solarplant-1 USA
Peak power [MWe]	0.1	0.4	4.9
Dish diameter [m]	4.8	7.0	24 × 1.5
Receiver fluid	diphyl	syltherm	water
Fluid temp [°C]	340	400	371
Aperture area [m²]	1025	4386	30261
Concentration ratio	210	235	~1500
Global efficiency [%]	9	14.5	
Operational	1981	1983	1984
Cost [$/W]			3.8

1.2.4.2.1 Sulaibiya, nr. Kuwait city, Kuwait

*** Collectors: 56 Messerschmitt-Boelkow-Blohm 18.3 sq.m dishes.

*** Comments: This is a stand-alone system designed to provide electric power to a remote agricultural research station. It is not grid connected.

*** Specific problems:

— Collector area under-designed as a result of inaccurate insolation data. (Only 70–80 kWe achieved compared to design figure of 100 kWe)
— Serious reflectance losses for collectors.
— Collectors washed every 4 weeks in summer but this is not considered sufficient.
— It was necessary to construct a 2.5 m high wall in order to protect collectors from sand storms.
— System can only be started up if insolation level exceeds 500 W/sq.m and shows an increasing trend.

*** Sources: [21–23].

1.2.4.2.2 "STEP", Shenandoah nr. Atlanta GA, USA

*** Collectors: 114 Solar Kinetics 38.6 sq.m dishes.

*** Comment: This system was designed to supply electricity, process steam and energy for cooling to a textile plant. The surplus electric power is sold to the local utility.

*** Specific problems:

— Energy output lower than expected.
— Flow balancing problems with the collectors.
— Various component problems.
— Control problems.

*** Sources: [24, 25].

1.2.4.2.3 "Solarplant-1", nr. Warner Springs CA, USA

*** Collectors: 700 LaJet Energy Company LEC-460 modules. (Each module comprising 24 parabolic dishes of 1.8 sq.m each)

*** Comment: No critical discussion of system performance has yet been published in the technical literature. A highly detailed, semi-popular, article appeared in SunWorld magazine [26], which is the source of the information quoted here. The reader should also see Prof. Roy's extended abstract in these proceedings.

*** Source: [26].

1.2.4.2.4 Economics of parabolic dish generating stations

The figures quoted in ref. [26] for the "Solarplant-1" system are impressive when compared with the economics of either the existing or projected central receiver projects. The total installation cost of "solarplant-1" is reported to have been $18.65 million for this 4.92 MWp system. No details are given regarding operation and maintenance costs or parasitic losses for this facility which is reported to have gone into operation in August 1984, so it is not possible to ascertain the true meaning of the quoted 2c/kWh for its power production.

1.2.4.3 *Distributed-field, reflection trough, solar power stations*

In Table 1.2.7 we list the six trough-type solar power stations that have been built to date, with peak energies equal to or greater

Table 1.2.7: Distributed trough power stations

	Vignola France	Meeka-Tharra Australia	Coolidge USA	DCS Spain	NIO-2 Japan	SEGS 1 USA
Peak power [MWe]	0.1	0.1	0.2	0.5	1.0	14.7
Tracking axes	1	2	1	1.2	2 + 0	1
Receiver fluid	gilotherm	transcal N	caloria	santotherm	water	Exxon 500
Fluid temp [°C]	250	290	288	295	370	307
Collection area [m²]	1,175	918	2,108	5,172	11,160	71,654
Concentration ratio	30	40	36	3,642	135	19
Global efficiency [%]	9	7.7	9.5	10.1	9.6	~19
Operational	1981–	1982–	1980–	1981–1984	1981–1984	1984–
Cost [$/W]		36	26			4.2

than 100 kWe. It should be pointed out that the troughs are not necessarily simple parabolas of cross-section: the Vignola plant, for example, employs a tracking linear receiver at the focus of a fixed (i.e. stationary), E-W axis, trough of effective circular cross-section. The Nio plant employs walls of tracking plane mirrors which direct the light on to fixed, E-W axis, parabolic troughs. Furthermore, a variety of tracking modes is employed for those systems which do employ parabolic troughs: Some systems employ horizontal collectors that track about a north-south axis, while others employ full two-axis tracking. Finally, at the time of going to press, the SEGS-1 entry in Table 1.2.7 is no longer the largest solar electricity generating facility in the world: The same company has since installed SEGS-2 at an adjacent site, with a peak output power of 30 MWe.

1.2.4.3.1   Vignola nr. Ajaccio, (Corsica), France

*** Collectors: 64 built-on-site 18 sq.m, E-W axis, non-tracking, segmented cylinder modules with tracking receiver. (constructed by the French Commissariat a l'energie atomique).

*** Specific problems:

— Reference [27] gives a detailed description of the facility and all its subsystems, but does not refer to any problems that may have arisen in its construction or operation.

1.2.4.3.2   Meekatharra, 650 Km north of Perth, Western Australia

*** Collectors: 30 Machinenfabrik-Augsburg-Nurnberg 30.6 sq.m, two-axis tracking, parabolic trough modules.

*** Comment: This is a hybrid system in which a screw expansion engine and generator are powered by heated oil, part of which comes from solar collectors and part from the waste heat recovery system of a conventional diesel power plant.

*** Specific problems:

— Excessive soot build-up on heat exchangers necessitated modification.

— Dual action of lube oil as screw sealant necessitates complicated oil-steam and oil-water separation sub-system. Sub-system
breakdown caused by mechanical failure of a shaft seal.
— Initial separate hot/cold tank storage judged relatively inflexible
to low operation temperatures and variable demand load. Storage
redesign under consideration.

*** Reference: [28].

1.2.4.3.3 Coolidge solar power plant, nr. Coolidge (Az), USA

*** Collectors: 384 Acurex 5.58 sq.m, N-S axis, tracking parabolic
troughs.

*** Comment: This system was designed to power an irrigation
project and to supply surplus generated electricity to the grid.

*** Specific problems:

— Simplification of collector piping, by removal of loop flow control
valves and buffer tank, reduced initially high energy losses.
— Rain caused delamination of reflective film.
— Fires broke out on two occasions as a result of leakage of the
(Caloria) heat transfer oil. Leak detection and stoppage deemed
to be high priority maintenance tasks.
— Various equipment problems caused by environmental factors.
— Various problems caused by equipment manufacture and instal-
lation shortcomings.

*** References: [29, 30].

1.2.4.3.4 IEA "DCS", Tabernas nr. Almeria, Spain

*** Collectors: 2,408 sq.m of Acurex Model 3001 parabolic troughs,
tracking about an E-W axis, plus. 2,496 sq.m of M.A.N. parabolic
troughs, employing two-axis tracking.

*** Specific problems:

— Power and water failures in this remote desert site.
— Available insolation was over-estimated at design stage.
— Shipping damage en-route (via rough roads) to site.

— Mis-installed and misconnected components.

— erratic electronic controls caused by grid voltage spikes.

*** References: [5, 31].

1.2.4.3.5   "PPM", Nio, (Shikoku) Japan

*** Collectors: 2480 Japanese-made 4.5 sq.m plane mirrors, employing two-axis tracking, in tandem with 124 non-tracking 13.68 sq.m parabolic troughs fixed along an E-W axis.

*** Comment: The PPM and CRT systems were designed to complement each other, CRT providing more power in summer and PPM more in winter.

*** Specific problems:

— Frictional ware on mirror tracking mechanism caused some problems.

— Bending of receiver tube in solar superheater led to 3% light loss.

*** References: [12, 32].

1.2.4.3.6   "SEGS-1", Daggett (CA), USA

*** Collectors: 4,480 Luz 16.0 sq.m parabolic troughs, tracking about a N-S axis.

*** Comment: An extensive discussion of this system by M. Margalit is reprinted in these proceedings.

*** References: [33–36].

1.2.4.3.7   Economics of reflecting-trough power stations

The cost per installed peak watt has dropped impressively from $26 for the Coolidge plant, installed in 1980 — I am ignoring the even more costly Meekatharra plant because it was only about 50% solar — to about $4 for SEGS-1 (and supposedly less for SEGS-2). This fact alone renders this kind of technology a leading contender in the struggle to achieve a solar alternative to fossil fuel. It must however be remembered that, at present, the economic attraction of this kind of system to certain utility companies does not reside in its ability to save fossil fuel.

1.2.5 *Photovoltaic systems*

The great attraction of photovoltaic cells as a source of solar generated electricity resides in their great simplicity: A photovoltaic station need not contain any moving parts. There is also no inertia in a photovoltaic system since the cells respond instantly to illumination. This point is well illustrated in Figs. 1.2.3 and 1.2.4 which show, respectively, the electric power output of the Lugo photovoltaic system and the Themis solar thermal facility.

In both examples sunrise is seen to occur at about 6:30 am, but whereas the Lugo system has reached maximum power output by about 10:00 am (having started to supply electricity at sunrise), Themis does not begin to feed into the grid until about 12:30 am.

Another attractive feature of photovoltaic power stations is the speed with which they can be installed. In fact, at the time this review was presented orally, I was able to list the existing photovoltaic facilities in a table similar to Tables 1.2.3, 1.2.6 and 1.2.7. Now (at the time of writing up the review), their number exceeds 20. Therefore

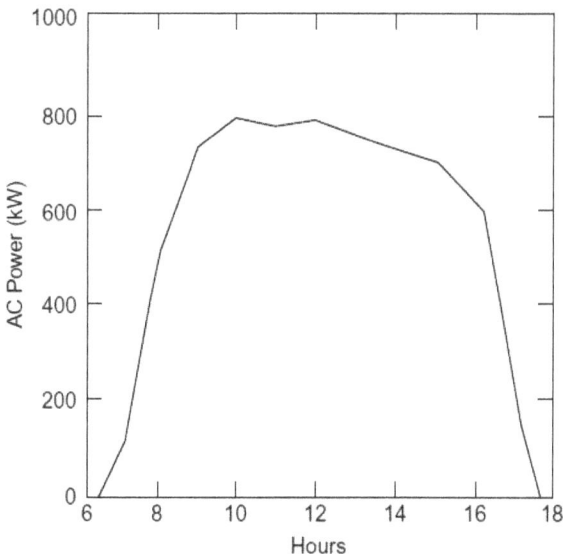

Figure 1.2.3: Output power of the Lugo photovoltaic system throughout a typical day

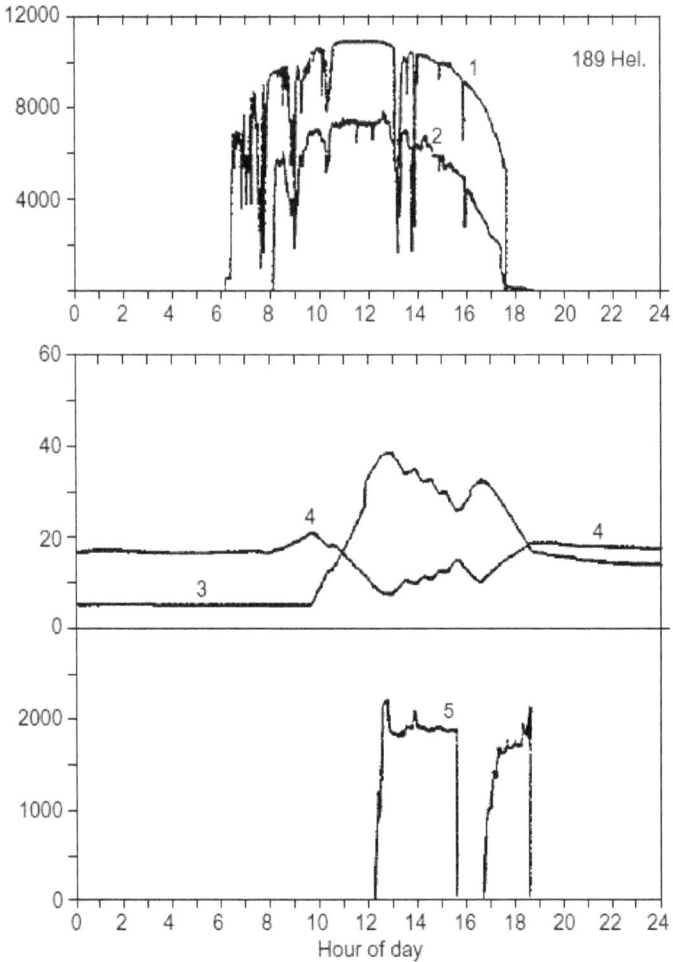

Figure 1.2.4:   Output power of the Themis solar-thermal system throughout a typical day

inclusion in a table is no longer feasible. Instead, I shall list them singly, drawing attention to the major features. Again, as with the solar thermal facilities, my emphasis will be on the negative aspects. It must be kept in mind however, that, in most cases, the systems have not been operating long enough for problems to have been fully documented.

1.2.5.1 *Bridges National Monument (UT), USA*

Installed in 1979, this was one of the first "large" photovoltaic power projects. The system comprises 1,700 sq.m of photovoltaic cells manufactured by three companies: Motorola, Arco Solar and Spectrolab. It supplies most of the electricity for a remote National Park which was previously served entirely by a diesel-powered generator. The system is reported to have cost $31/Wp.

References: [37, 38].

1.2.5.2 *Kythnos Island, Greece*

System comprises 800 Siemens type SM 144-9 modules. Each module consists of 144, 10 cm diameter, monocrystalline silicon cells. The array has an area of 1,200 sq.m and a nominal rating of 100 kWp. The system entered operation in 1982.

References: [39, 40].

1.2.5.3 *Hamamatzu City, (Shizuoka Pref.) Japan*

A 100 kWp facility that provides the initial charging of batteries produced by an automobile battery factory.

References: [41, 42].

1.2.5.4 *San Augustin de Guadalix, nr. Madrid, Spain*

A 100 kWp central power station based, largely, on bi-facial singe-crystal cells manufactured by the Isofoton corporation. All modules are stationary, flat-plate types and the entire facility occupies 7,000 sq.m.

Reference: [43].

1.2.5.5 *Puerto Rico*

A 100 kWp stationary flat-plate facility.

1.2.5.6   *Caltrans, California, USA*

A 100 kWp stand-alone system which complements three existing
diesel generators of the California Department of Transport. Modules
are Arco Solar "square" single-crystal cells.

Reference: [44].

1.2.5.7   *Shopping center, Lovington (NM), USA*

A 104 kWp flat-plate array. The system is reported to have cost
$30/Wp.

Reference: [38].

1.2.5.8   *Oklahoma center, Oklahoma city (OK), USA*

System consists of 1,512 Solarex 65 cm × 128.5 cm modules with a
nominal rating of 106 kWp but augmented to 140 kWp by stationary
plane mirrors. Each module contains 72 polycrystalline 10 cm × 10 cm
cells.

System went into operation in 1982, surplus electricity is sold to
the utility. Total system cost was $2.98 M. (about $21/Wp). Output
power is lower than expected: low module output is suspected of
being the cause.

References: [38, 45].

1.2.5.9   *John Long 24-house project, Phoenix (AZ), USA*

A housing project supplied by a 192 kWp stationary array of Arco
Solar "square" single-crystal photovoltaic cells. The system is grid —
connected to the local utility, which purchases surplus power in the
following way.

The utility's buying-selling prices are: 0.0503–0.0743 $/kWh
during the period 12:00 to 22:00 and 0.0265–0.0293 $/kWh during
off-peak hours.

The system is reported to have cost about $10/Wp.

References: [44, 46].

1.2.5.10   *Tsukuba University, (Ibaraki Pref.), Japan*

A 200 kWp array supplying electricity for lighting and laboratory equipment.

References: [41, 42].

1.2.5.11   *Ichihara city, (Chiba Pref.), Japan*

A grid-connected, distributed system comprising a total of 200 kWp of cells on rooftops and other available space.

References: [41, 42].

1.2.5.12   *Solarex factory, Frederick (MD), USA*

2,500 sq.m of roof-mounted Solarex polycrystalline cells (nominal rating: 200 kWp) supply all the electricity requirements for this solar cell manufacturing "breeder" plant.

1.2.5.13   *Sky Harbor Airport, nr. Phoenix (AZ), USA*

This system consists of 80 two-axis trackers. Each tracker comprises 272 individual solar cells each placed at the focus of an R = 33x fresnel lens. The fresnel lenses have apertures of approximately 30 cm × 30 cm and the power rating of the entire system is 225 kWp. The trackers were built by Martin Marietta Corp. The system has been operational since 1982 and is reported to have cost \$11/Wp.

References: [38, 47, 48].

1.2.5.14   *Pellworm Island, West Germany*

The system consists of a (mainly) stationary flat-plate array of 17,568 AEG-Telefunken type PQ 10/20/0 modules. Each module contains 20 multi-crystalline cells of dimensions 10 cm × 10 cm. The nominal power rating is 300 kWp. The system was installed in 1983 at a reported cost of about \$15.6/Wp. It occupies a total land area of 16,500 sq.m.

References: [40, 49, 50].

1.2.5.15   *Georgetown Univ. Intercultural Center, Washington DC, USA*

This system consists of stationary, architecturally integrated, flat plate modules of nominal power 300 kWp. The modules contain Solarex microcrystalline cells and occupy about 3,300 sq.m. The system became operational in 1984. It is reported to have cost $21/Wp.

Reference: [38].

1.2.5.16   *Mississippi county community college, Blythesville (AK), USA*

System consists of 270, N-S axis, parabolic trough collectors made by Solar Kinetics Corp, each having an aperture area of 13 sq.m and a concentration ratio of about R = 42×. The collectors concentrate light onto linear modules of single-crystal photovoltaic cells made by Solarex. The cell modules are mounted on a liquid-cooled heat sink which enables the system to supply about 22 GJ of thermal energy per day (at peak output), in addition to 320 kWp of electrical power. The photovoltaic cells are 2.5 cm × 5.0 cm in size, and operate with an efficiency of about 12% at 55°C.

The system became operational in 1981 but the electricity-generating part could not be made to function in a trouble-free manner. It is reported to have cost $30/Wp. The last published DOE summary on the project (dated October 1983) mentioned investigations that were underway to convert the system from a photovoltaic to a thermal facility.

References: [38, 51, 52].

1.2.5.17   *City of Austin project, (TX), USA*

A 300 kWp project of the local utility, for city power generation.

Reference: [44].

1.2.5.18   *The Soleras project, Saudi Arabia*

This system is similar to that at Sky Harbor airport in the USA, but larger. Its rated power output is 350 kWp. There are 160 arrays, each containing 32 cell modules. The project, which is well-documented, was installed in 1981 and the only serious problem appears to have been the gradual failure (as of August 1984) of 6–7% of the modules. Module degradation has been attributed to manufacturing faults rather than environmental effects. The system is reported to have cost $53/Wp.

References: [38, 53–55].

1.2.5.19 *"Lugo" power station, nr. Hesperia (CA), USA*

This system comprises 108 two-axis trackers each containing 256 Arco Solar flat-plate modules of "round" single-crystal photovoltaic cells. The project occupies a land area of 20 acres and has a nominal power rating of 1 MW. Reference 56 provides a valuable indication of the various sources of power loss associated with this plant. According to Arnault *et al.* the conversion of the dc output to ac involves a total loss of 18%, broken down as follows:

Module/panel interconnects:	0.7%
Field wiring:	1.3%
Inverter and switching gear:	4.0%
Power for tracking:	3.0%
Shadowing:	2.5%
Module mismatch:	3.0%
Nighttime ac standby:	2.0%
Operational downtime:	1.5%

References: [56–59].

1.2.5.20 *Centralized power station, Saijo (Ehime Pref.), Japan*

This 1 MWp system consists of stationary, flat-plate modules made by various photovoltaic cell manufacturers. It started to produce 200 kWp in 1982 and was enlarged step-by-step to 1,000 kWp in 1986. According to Ref. [42], "It is difficult to find a wide space for such a centralized plant in Japan."

References: [41, 42].

1.2.5.21   *S.M.U.D., Rancho Seco nr. Sacramento (CA), USA*

This central power facility, operated by the Sacramento Municipal Utility District, comprises two subsystems each of nominal power 1.2 MWp. The arrays are Arco Solar single-crystal modules mounted on Acurex single (N-S) axis trackers. The two subsystems constitute the first two phases of what was planned to be a 100 MWp facility, but for the time being (DOE summary, October 1985) all subsequent phases have been cancelled. The present system is reported to have cost $10/Wp.

References: [38, 44].

1.2.5.22   *Carrisa plain facility, nr. Los Angeles (CA), USA*

Currently the largest photovoltaic facility, with a power rating of 6.5 MWp, this system is similar in principle to Lugo, 90% of the trackers being of the same (33ft × 36ft) aperture area but enhanced by plane side mirrors set at an angle of 120 deg to the plane of the cell modules. The remaining 10% of the trackers are not mirror-enhanced but merely larger (40ft × 40ft). Carrisa Plain employs Arco Solar's "square" single-crystal cells.

References: [57, 60].

1.2.6   *Conclusions*

In the above pages I have tried to collect information about most, if not all, of the solar electricity-generating facilities in the world, with a power rating above 100 kWp. I have tried to concentrate on the problems rather than the success part of the story in order to emphasize how much there is that we can learn from what has already been done by others. A few general conclusions may already be drawn by perusing this summary.

First, the problem of keeping collectors clean is ubiquitous. In desert countries such as Israel a prime consideration will have to be the availability of water for washing the collectors — whatever kind they will be. It should not be forgotten that a 1,000 MWp solar electric power station would involve about 5 sq.km of solar collectors — which is a lot of area to wash every week or so.

Second, nearly all of the solar thermal plants suffered some form of leakage problem. If the leaking fluid is heat transfer oil there is a potential fire hazard, and indeed, fires did occur (See section 1.2.4.3.3). It is of the utmost importance that all solar thermal systems should have fail-safe leak detection, flow-stop and defocusing devices — particularly if the prevailing winds are in the direction of inhabited areas, as is the case for our settlement at Midreshet Sede Boqer.

A third problem to which remote locations are particularly prone (See section 1.2.4.3.4) involves electric power and water failures. All safety systems must be designed to accommodate such eventualities.

Fourth, when it comes to central power stations, the Japanese already have a problem with finding suitable land area to accommodate the collectors. Dr. Horigome's comment quoted in section 1.5.1.20 vis-a-vis the Saijo facility should serve as a timely warning for small countries such as ours. The Negev is certainly large by Israeli land standards but our future solar power stations should be as small as possible. In Table 1.2.8, I have prepared a comparative land usage display for the largest power stations that currently exist. Dividing the land area by the total peak power output of the facility one sees that current installations range from 56 kWp per dunam*

Table 1.2.8: Land requirements for solar electric power stations

Station	Aperture [m²]	Peak power [MW]	Land area [m²]	Land utility [%]	Land output [kW/dunam*]
SEGS 1	72,000	14.7	264,000	27	56
Solarplant 1	30,000	4.9	160,000	19	31
Solar 1	72,500	12.1	520,000	14	23
Hesperia	10,000	1.0	80,000	12	12
Beit Ha'Aravah	250,000	5.0	400,000	62	12

[*1 dunam = 0.1 hectare]

Figure 1.2.5:   Map of the Negev showing 4 schematic 1000 MWp solar plants for scale (white rectangles): three land-based plants, scaled according to SEGS-1; and one Dead Sea floating pond, scaled according to the Beit Ha'Aravah plant [Underlying map courtesy the U.S. Central Intelligence Agency]

(for SEGS-1) to 12 kWp per dunam* (for Lugo — and even less for Carrisa Plain). Thus so far, Luz seem to be able to utilize the land most efficiently. [*Editor's comment:1 dunam = 0.1 hectare].

Figure 1.2.5 shows a map of the Negev with three 1,000 MWp power plants, sized according to the land requirements of SEGS-1, and one floating 1,000 MWp pond in the Dead Sea, sized according to the Beit Ha'Aravah facility. Such a quartet of solar plants could provide comparable generating capacity to that which currently exists in Israel.

Fifth, given that there may be a potential land problem, the most flexible technology for "spreading around" is clearly photovoltaics. For factories (sections 1.2.5.1.3 and 1.2.5.1.12), shopping malls (section 1.2.5.1.7), private houses (section 1.2.5.1.9), etc., this point is clear. Indeed the mutual financial arrangement between the Phoenix homeowners and their local utility, as summarized in section 1.2.5.1.9, (and other arrangements in the U.S.A.) should serve as a basis for study in countries where such systems are under consideration.

Sixth, for large-scale central power generation, solar-thermal appears to be ahead on three counts: land utilization (already emphasized), cost (by a factor of two to three) and efficiency (20% compared with 10% for photovoltaics). But the situation could change. Photovoltaic efficiencies could double and prices drop, for example, in the manner mentioned by Dr. Berman in his contribution to this symposium. Furthermore, the S.M.U.D. use of photovoltaic modules mounted on N-S axis trackers should be similarly efficient (from the land usage viewpoint) to Luz's at SEGS-1. And, as we have shown in Table 1.2.1, the Negev would supply a substantial insolation bonus to non-concentrating PV modules mounted on such trackers, compared with the loss of diffuse radiation associated with parabolic trough concentrators such as Luz's.

Finally, important as it is to keep abreast of current solar technologies from various parts of the world, test facilities such as ours should also serve as research facilities for fostering new ideas. In particular, they can serve as a neutral ground where component specialists from different areas can get together to work on new kinds of system. Typical examples of potential joint ventures might be: The coupling of high-efficiency solar thermal collectors with a magneto hydrodynamic generator; the coupling of a low-temperature organic Rankine generator with a non-saline solar pond; the coupling of photovoltaic cells with reflectors based on recent developments in optics, and generally, a testing ground for the ideas that abound in the various laboratories around the globe.

1.2.7 *Acknowledgements*

I wish to thank Dr. Avraham Zangvil for allowing me to use his raw insolation data for Sede Boqer, prior to publication, in order to construct Table 1.2.1, and Dr. Amos Zemel for help in this construction. My thanks are also due to Prof. Yossi Applebaum and Dr. Jeff Gordon for valuable help in locating some of the listed references.

References

1. N.R. Sheridan, "Rankine cycle energy converters for solar ponds" in *Solar World Congress* ed. S.V. Szokolay (Pergamon, Oxford 1984) p. 1473.
2. R. Winston, "Light collection within the framework of geometrical optics", *J. Opt. Soc. Am.* 60 (1970) 245.
3. F. Pharabod, "The thermo-electric conversion of solar energy", in *Intersol* 85 ed. E. Bilgen and K.G.T. Hollands (Pergamon, New York, 1986) p. 1379.
4. B. Floss and D. Stahl, "The 500 kWe CRS solar power plant in Almeria, Spain: Design, construction and commissioning experience" in *Solar World Congress* (*Op. cit.*) p. 1522.
5. W. Grasse, "Operational experiences with the SSPS project of the IEA in Almeria, Spain" in *Solar World Congress* (*Op. cit.*) p. 1450.
6. W. Grasse and M. Becker, "Central receiver system (CRS) in the small solar power systems project (SSPS) of the International Energy Agency (IEA)", *ASME J. Solar Energy Engineering* 106 (1984) 59.
7. G. Cefaratti and J. Gretz, "Eurelios", *Sunworld* 5 (1981) 104.
8. D. Borgese, C. Corvi, J. Gretz and A. Strub, "Eurelios: The European Community 1 MW (e1) solar tower plant at Adrano (Italy), description and preliminary results" in *Solar World Congress* (*Op. cit.*) p. 1577.
9. D. Borgese, G. Dinelli, J.J. Faure, J. Gretz and G. Schober, "Eurelios, the 1-MW(el) heliocentric power plant of the European Community program", *ASME J. Solar Energy Engineering* 106 (1984) 66.
10. C. Corvi, G. Dinelli, J. Gretz and A. Strubb, "Operating experience of a demonstration solar central receiver power plant" in *Intersol* 85 (*Op. cit.*) p. 1413.
11. T. Hirono and T. Horigome, "A 1-MWe central receiver type solar thermal electric power pilot plant", *ASME J. Solar Energy Engineering* 106 (1984) 90.
12. S. Wakamatsu, N. Ikeda, T. Horigome, T. Saishoji, K. Fukuda and I. Sumita, "Operational research of 1 MWe solar thermal electric power plants" in *Intersol* 85 (*Op. cit.*) p. 1419.

13. A. Munoz Torralbo, C. Hernandez Gonzalvez, C. Ortiz Roses, J. Avellaner Lacal and F. Sanchez, "A Spanish 'Power Tower' solar system: Project CESA-1", *ASME J. Solar Energy Engineering* 106 (1984) 78.

14. L.P. Drouot and M.J. Hillairet, "The Themis program and the 2500-kW Themis solar power station at Targasonne", *ASME J. Solar Energy Engineering* 106 (1984) 83.

15. J.J. Bezian, "Themis solar power plant first evaluation results", in *Intersol* 85 (*Op. cit.*) p. 1408.

16. R.L. Gervais and R.W. Hallet, "Design and operation of Solar One: Application to commercial systems", in *Solar World Congress* (*Op. cit.*) p. 1544.

17. J.J. Bartel and P.E. Skvarna, "10-MWe solar thermal central receiver pilot plant", *ASME J. Solar Energy Engineering* 106 (1984) 50.

18. J.R. Roland and K.M. Ross, "Solar central receiver technology development and economics — 100 MW utility plant conceptual engineering study" in *Energy Technology X* ed R.F. Hill, (Government Institutes Inc, Rockville MD, 1983) p. 1421.

19. Solar Engineering & Contracting, January/February 1985, p. 7.

20. A.T. Marriott, "Solar electric power from parabolic dishes" in *Solar World Congress* (*Op. cit.*) p. 1438.

21. S. Moustafa, W. Hoefler, H. El-Mansy, A. Kamal and D. Jarrar, "Design specifications and application of a 100 kWe (700 kWth) cogeneration solar power plant", *Solar Energy* 32 (1984) 263.

22. H. Zewen, G. Schmidt and S. Moustafa, "The Kuwait solar thermal power station: Operational experiences with the station and the agricultural application" in *Solar World Congress* (*Op. cit.*) p. 1527.

23. S.M.A. Moustafa, H. El-Mansy, A. Elimam and H. Zewen, "Operational strategies for Kuwait's 100 kWe/0.7 MWth solar power plant", *Solar Energy* 34 (1985) p. 231.

24. R.W. Hunke, W.S. Mertz and A.J. Poche, "Definitive design of the solar total energy large scale experiment at Shenandoah, Georgia" in *Sun II* eds. K.W. Boer and B.H. Glenn (Pergamon, New York, 1979) p. 1116.

25. W.B. Stine, E.J. Ney, A.A. Heckes and J.M. Connolly, "Performance and operating experience of the solar total energy project at Shenandoah, Georgia" in *Intersol* 85 (*Op. cit.*) p. 1398.

26. D. Carroll, "Solarplant 1", *SunWorld* 9 (1985) 10.

27. J.L. Boy-Marcotte, M. Dancette, J. Bliaux, E. Bacconnet and J. Malherbe, "Construction of a 100 kW solar thermal-electric experimental plant", *ASME J. Solar Energy Engineering* 107 (1985) 196.

28. G. Hanselmann, E. Hellweg, T.S. Crawford and V.F. Power, "Combined solar and waste heat utilisation in a power plant at Meekatharra, Western Austraila", in *Solar World Congress* (*Op. cit.*) p. 1598.

29. D.L. Larson, "Coolidge 150 kWe solar irrigation project" Arizona Agricultural Experiment Station Publication on No. 366 (1980).

30. D.L. Larson, "Coolidge solar power plant: Operational results", in *Solar World Congress* (*Op. cit.*) p. 1828.

31. F.A. Schraub and H. Dehne, "Electric generation system design: Management, startup, and operation of IEA distributed collector solar system in Almeria, Spain", *Solar Energy* 31 (1983) 351.

32. M. Chinen, M. Tsukamoto, I. Sumida, S. Anzai, T. Taki, S. Sato and S. Sasaki," Performance characteristics of solar thermal power generation system with flat plate mirror and parabolic mirror", in *Solar World Congess* (*Op. cit.*) p. 1581.

33. D. Kearney, I. Kroizer, C.E. Miller and D. Steele, "Design and preliminary performance of Solar Electric Generating Station 1 at Daggett, California", extended abstract in *Intersol* 85 (book of abstracts) p. 312.

34. S. Mor, Y. Haratz, Y. Sarel, I. Kroizer and D. Kearney, "Design and performance of an advanced parabolic trough collector", extended abstract in *Intersol* 85 (book of abstracts) p. 317.

35. W. Joseph and J. Hogsett, "Design and preliminary performance of a two-tank oil storage system at SEGS1", extended abstract in *Intersol* 85 (book of abstracts) p. 340.

36. D. Kearney, I. Kroizer, C.E. Miller and D. Steele, "Design and preliminary performance of Solar Electric Generating Station 1 at Daggett, California", in *Intersol* 85 (*Op. cit.*) p. 1424.

37. E.F. Lyon, "A 100 kW peak photovoltaic power system for the Natural Bridges national monument", in *Sun II* (*Op. cit.*) p. 1732.

38. L.M. Magid, "U.S. photovoltaic systems experience: Prospects for the future", in *Proceedings of the 17th IEEE Photovoltaic Specialists Conference, Kissimmee, FL, May 1984.*

39. ———, "Kythnos photovoltaic power plant", in *Photovoltaic Power Generation*, Series C, vol. 1 (Proc. of final design review meeting on the European Community photovoltaic pilot projects, Bruxelles, 30 Nov–2 Dec, 1981) p. 217.

40. F.C. Treble, "The CEC photovoltaic pilot projects", in *Proceedings of the 6th European Community Photovoltaic Solar Energy Conference*, London, April 1985.

41. K. Kurokawa, H. Akabane, H. Hosokawa and K. Murakami, "A series of R&D on solar photovoltaic conversion systems of Japan", in *Solar World Forum* eds. D.O. Hall and J. Morton (Pergamon, Oxford 1982) p. 2755.

42. T. Horigome, "Solar energy development in Japan" (15 pp typewritten report, dated March 1986), private communication from the author.

43. J.S. Solera *et al.*, "100 kW experimental photovoltaic power plant", in *Proceedings of the 6th E.C. Photovoltaic Solar Energy Conference.* (*Op. cit*).

44. G.J. Shushnar and R. Yenamandra, "Trends in recent applications of large photovoltaic systems", in *Proceedings of the ASME Solar Energy Conference*, Anaheim, CA, April 1986, p. 184.

45. Y.P. Gupta, "Oklahoma Center for Science and Arts: Intermediate photovoltaic system application experiment, Phase II — Final Report" (DOE/ET/20630-T1) January 1984.

46. S. Andrews, "A solar developer breaks new ground", *Solar Age* 11 (Dec 1985/Jan 1986) p. 40.

47. L.O. Herwig, "Technology and economic status of concentrating photovoltaic systems in the United States", in *Solar World Congress* (*Op. cit.*) p. 1565.

48. W.J. McGuirk, "Fabrication and construction of the Sky Harbor Airport solar photovoltaic concentrator project, Phase II — Progress Report from March 11, 1980–June 30, 1982" (DOE/ET/20624-1.60) January 1984.

49. _____, "300 kW photovoltaic pilot plant Pellworm", in *Photovoltaic Power Generation* (*Op. cit.*) p. 179.

50. H.J. Lowalt and B. Proetel, "The 300 kW Pellworm solar power station: Performance and experience", in *Proc. of the 6th E.C. Photovoltaic Solar Energy Conference.* (*Op. cit.*).

51. E.M. Henry and H.V. Smith, "Mississippi County Community College solar photovoltaic total energy project", in *Sun II* (*Op. cit.*) p. 1729.

52. H.V. Smith, "The Mississippi County Community College large-scale demonstration project: A success story", in *Proceedings of the 4th European Community Photovoltaic Energy Conference*, 10–14 May 1982, Stresa, Italy, p. 94.

53. F. Huraib, B. Khoshaim, A. Al-Sana, M.S. Imamura and A.A. Salim, "Design, installation and initial performance of 350-kW photovoltaic power system for Saudi Arabian villages", in *Proc. of 4th E-C Photovoltaic Solar Energy Conference* (*Op. cit.*) p. 57.

54. _____, "SOLERAS photovoltaic power systems project: Final report; photovoltaic power seminar February 20–23 1983" (Midwest Research Institute, Kansas City, report No. MRI/SOL 0102).

55. F.S. Huraib, M.S. Imamura, A.A. Salim and N. Rao, "SOLERAS photovoltaic power systems project: Interim report; module failure analysis" (Midwest Research Institute, Kansas City, report No. MRI/SOL 0101) October 1984.

56. R.J. Arnault, E. Berman, C.F. Gay, R.E.L. Tolbert and J.W. Yerkes, "The ASI one-megawatt photovoltaic power plant", in *Solar World Congress* (*Op. cit.*) p. 1624.

57. J.C. Arnett and R.F. Reinoehl, "Installation and performance of Lugo and Carrisa Plain photovoltaic systems", in *Proceedings of the 166th Electrochemical Society Meeting*, October 1984.

58. N.W. Patapoff and D.R. Mattijetz, "Utility experience with an operating central station MW-sized photovoltaic plant", in *IEEE Transactions on Power Apparatus and Systems*, August 1985, p. 2020.

59. N.W. Patapoff Jr, "Two years of inter-connection experience with the 1 MW at Lugo", in *Proceedings of the 18th IEEE Photovoltaics Specialists Conference*, Las Vegas, NV, October 1985.

60. T. Hoff and G. Shushnar, "Two years of performance data for the world's largest photovoltaic power plant", in *Proceedings of the IEEE Power Engineering Society Meeting*, Summer 1986.

1.3   Editorial Insert on Solar Ponds (2019)

As it happened, this was one of the solar power technologies that, although technologically successful at the time, did not develop further owing to the perceived economic value that resulted from the experimental findings of the 1980s. Therefore, a few introductory words may be helpful to enable present-day readers to follow the relatively terse presentation of Benny Doron on the performance of the Beit Ha'Aravah plant, and the discussion that followed.

The idea was to use water at a temperature slightly below 100°C as an energy source for a specially designed turbine that was driven by the vapor from a low boiling-point organic fluid rather than by steam.

Ormat Corporation originally developed such turbines with the idea of using flat-plate solar water-heating panels as their heat source. However, when the latter turned out not to be economically feasible for this task, the company sought a far less-costly kind of solar water heater. This was the motivation for developing the salt-gradient solar pond, some naturally occurring examples of which had previously been observed in a number of locations around the world [Fig. 1.3.1].

The company consequently engineered an artificial pond whose thermal properties could be maintained in a quasi-stable state by mechanical means. To do this, what was essentially a low-cost hole in the ground, was dug and lined with water-proof plastic. It was then slowly filled with saline water to a depth of a few meters, starting with saturated salt solution at the bottom, and gradually decreasing in salinity until at the surface the pond water had close to zero salinity. This artificially-induced salinity gradient was maintained, against the force of natural diffusion, by flushing the surface of the pond with fresh water, and re-introducing salt into the pond's lowest level.

The pond was shallow enough to enable solar radiation to penetrate through to the bottom where it was absorbed, thus heating the underlying soil. The hot ground then heated the highly saline layer with which it was in contact, but the resulting hot concentrated brine was unable to rise, owing to its greater density than the less saline layers above. In essence, the pond's density gradient restricts natural convection to lie within two narrow layers: one close to

Figure 1.3.1:   A naturally occurring solar pond in a Red Sea cove slightly south of Eilat. Winter waves throw sea water into the cove, forming a shallow pond which gets cut off from the rest of the sea during the calmer summer season. The gradually evaporating pond water creates a salinity gradient and a resulting temperature gradient of several tens of degrees [Photo: D. Faiman]

its floor — a so-called "lower convective zone" — which can reach temperatures close to 100°C, because its saturated brine, although hot, is still too dense to rise and mix with the upper layers; and a so-called "upper convective zone" close to the surface, which is essentially at ambient temperature. Between these two convective zones, there is no convection: only constantly decreasing salinity and temperature gradients with decreasing depth.

The solar-heated brine is pumped out of the pond and through the input heat exchanger of the turbine. The out-flowing cooler brine is then returned to the pond at an appropriate level that does not disturb the salinity gradient. Figure 1.3.2 shows a schematic of such a system.

Ormat Corporation constructed two experimental systems: one at the 100 kilowatt scale in Ein Bokek; and a second at the 5 megawatt scale at Beit Ha'Aravah [Fig. 1.3.3],

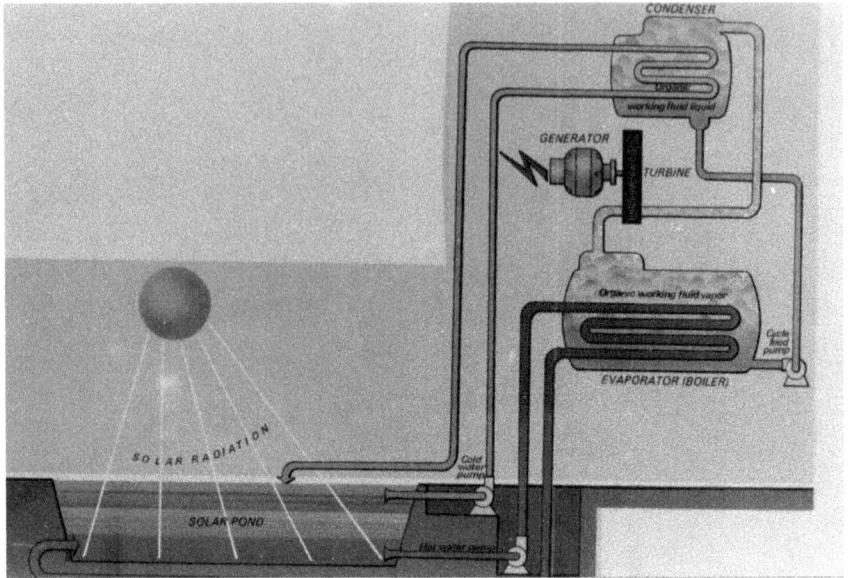

Figure 1.3.2:   Schematic of a solar pond power plant [Courtesy Ormat Corp.]

Figure 1.3.3:   The Beit Ha'Aravah 5 MW solar pond power plant [Photo: D. Faiman]

1.4 Summary of the First Year of Operation of the 5 MW Solar Pond Project at Beit Ha'Aravah, Israel

A presentation by Eng. Benny Doron (Ormat Turbines Ltd, Yavneh, Israel) at the First Sede Boqer Symposium on Solar Electricity Production, 23–24 February, 1986. [English translation by David Faiman, authorized by Mr. Doron.]

Abstract

After the successful operation of the solar pond at Ein Bokek with its power unit of 150 kW (in the month of December 1979) it was decided a 5 MW project at Beit Ha'Aravah whose purpose was to demonstrate a solar pond as a source of power on a large scale be set up.

Within the framework of the project we obtained data for the following issues:

— Operation on a large scale
— Cost of construction
— Maintenance on a large scale
— Electrical output

1.4.1 *Introduction*

In order to demonstrate a worthwhile power unit of industrial strength within the budgetary framework that was allocated to the project, it was decided that a 5 MW unit that would be powered by a 250 dunam [250,000 m^2] pond during only a few hours each day be set up.

At a later stage, owing to budgetary constraints, it was decided to install only half of the heat exchangers and not to install the cooling system for the condensers. This cutback enabled us to allow sufficient budget for experimentation and operation of the solar pond at full power, but at the cost of partial operation of the power plant and reduced efficiency for the conversion of heat to electricity.

1.4.2 *Performance of the ponds during the year 1985*

Efficiency of the pond is still rising on account of diminishing heat losses to the ground. During the year 1986 we will reach the expected

Table 1.4.1:   Expected and measured pond energies

		Expected	Measured
Solar radiation	W m^{-2}	215	205
Radiation reaching storage	W m^{-2}	47.3	40.4
Radiation reaching storage	%	22	19.7
Average annual heat production	W m^{-2}	34.4–38.7	31.6
Efficiency of the pond as a collector	%	16–18	15.4

efficiency. In addition an increase in radiation will affect the efficiency of the pond.

We have at hand a method to improve the transparency of the pond, which we shall employ in the coming year on the 40 dunam pond [40,000 m^2]. We expect a rise in efficiency to 20–22%.

1.4.3   *Cost of construction*

In spite of a cost of approximately US$ 24 per square meter of pond (including systems) at the start of the project, we (eventually) reached approximately US$ 10/m^2, which is the lowest cost collector-and-storage on the market.

Capital cost per kWh thermal is 0.34 US¢, assuming a radiation of 210 W/m^2 and a collection efficiency of 16% (at a temperature difference of 58 K) and capital return of 10%.

1.4.4   *Maintenance*

1.4.4.1   *Manpower*

Today, a single technician maintains the ponds and the pond-related systems. The same technician could also maintain a 1 km^2 pond. For safety reasons, an assistant technician is also on hand.

1.4.4.2   *Water*

Flushing during the year 1985 amounted to approximately 2.9 m^3 per m^2. For flushing, we use local sources of brackish water that have no alternative use.

1.4.4.3 *Chemicals*

During the year 1985 we used chemicals for water treatment, to the extent of approximately US$ 0.10 per m^2, or approximately 0.03 US¢ per kWh thermal.

1.4.5 *Power plant*

We examined the performance of the power unit a number of times, at its designed operating point, each time for a limited number of hours (as long as the cooling water did not heat up). Here are some typical measured results:

Table 1.4.2: Expected and measured performance of power unit

	Design point	Measured
Temperature of the pumped brine — Deg C	85	85.2
Temperature of the pumped cooling water — Deg C	28	28.5
Flow rate of brine — m^3 hr^{-1}	5000	4950
Flow rate of cooling water — m^3 hr^{-1}	5000	5070
Heat supplied to power unit — M kCal hr^{-1}	30.65	30.72
Heat exiting power unit — M kCal hr^{-1}	28.25	28.29
Gross electricity production — kW	2100	2150

In spite of the fact that performance of the power unit and the ponds were as expected, because the system was incomplete, a given amount of heat yielded a significantly lower [net] electrical output. Two main causes led to a decrease in net output — High temperature cooling water, and high internal [electricity] usage. Instead of an electrical output of 5000 kW and an internal requirement of up to 1500 kW, we worked at [output] powers of 1.0–1.2 MW with an internal requirement of approximately 0.8 MW.

The net electrical output per square meter was 2.65 kWh compared with an expected 12 kWh. This was the penalty incurred for having saved on the heat exchanger and the cooling system.

1.4.6 *Lessons*

From the technological standpoint it was proved that the system operates according to expectations, and there were even some pleasant surprises, particularly regarding maintenance. Nevertheless, for future systems we shall use more advanced power units, which have already been developed by Ormat: modular units as installed at geothermal wells in the USA. Such units are each 0.8–1.2 MW with an accompanying system for control and monitoring, and attachment points for the brine and water cooling lines.

Such units reduce pumping power (by virtue of the closer proximity of the power units to the pond), allow flexibility for operating the pond to supply baseload or peak demand, via optimization of the system for every kind of demand. This kind of setup allows for both normal operation and changes in demand during electricity supply, while maintaining a high capacity factor.

1.5 Discussion Following Benny Doron's Presentation

Q: For how many hours can one supply 5 MW from the Beit Ha'Aravah pond during the months December–January?

Doron: The Beit Ha'Aravah pond, which is relatively shallow (2.5 m in depth), will be able to supply 5 MW of electricity for 2–3 hours a day in mid-winter. In summer the period will expand to 12–14 hours a day. This is a ratio of 1:6. If the pond were deeper, e.g., 5–10 m, it would be possible to supply a constant daily peak demand throughout the year.

Q: To what extent does the fact that the pond is still in its transient stage affect the 15–16% efficiency that you observed during the first year?

Doron: To a great extent. During the first year the rate of heat loss to the ground amounts to about 8–10 W/m^2 out of 37–40 W/m^2 that reaches the storage layer. This rate of heat loss is expected to fall to about 1.5–2 W/m^2.

Q: But what about the time it takes to charge the pond itself?

Doron: It is quite short, but naturally depends upon the season you start the charging. We started during the last part of the year. Because of this the Beit Ha'Aravah pond only reached 69°C in December. The temperature remained at this level during January, and started to rise around the second week of February. On the other hand, in summer the pond heats up at 0.7–0.8 K per day, so that it does not require more than 2–3 weeks in order to achieve its final operating temperature. The short time needed for charging depends, of course, on the absence of any unusual sources of heat loss. For example, due to the rate of heat loss during the first year (about $10\,\text{W/m}^2$), the charging rate is longer, because we are heating the ground beneath the pond to a depth of tens of meters.

Q: Heat loss to the ground can be greatly affected by the flow of underground water. Is the site where your ponds are located particularly free from this phenomenon?

Doron: No. In fact we encountered underground water during our excavation for the Beit Ha'Aravah pond. However, I have to emphasize that the flow rate was [of order] meters or tens of meters per year. The heat loss this causes to the Beit Ha'Aravah pond — whose length is 700 m — is negligible. The effect was considerable at our much smaller Ein Bokek pond. At that pond — whose length is only 70 m — the annual efficiency that was achieved was only 9%.

1.6 Editorial Comment (2019)

An instructive technical comment may be drawn from Mr. Doron's Table 1.4.2. Namely, a gross electrical power output of 2.1 MW when compared with a thermal power input of 30.72 M kCal hr^{-1} (=35.7 MW) implies a turbine efficiency of 5.9%. This approaches encouragingly near to the Curzon–Ahlbor efficiency $[1 - \sqrt{(301.5/358.2)}] = 8.2\%$ of an ideal Carnot engine operating at its maximum power point between temperatures of 85.2 and 28.5°C, indicating that by 1986 a high level of maturity had already been reached for this kind of technology.

The Beit Ha'Aravah solar pond system was operated successfully for a number of years, albeit without the hoped — for additional heat exchangers and cooling system. It was eventually shut down after a government conclusion that the technology was not cost-effective. It should be mentioned that the project was originally funded during a period of rampant inflation: Budget proposals being calculated in US dollars, but funding being provided after significant delay, in Israeli Shekels. This was the extreme budgetary constraint that Mr. Doron alluded to. Fortunately the Ormat company was subsequently able to achieve commercial success for these turbines as generators of geothermal power in various parts of the world. However, whether large-scale, solar ponds might yet provide cost-effective electric power remains an open question. The *unique* attraction of this technology, compared to all other renewable energy systems, is that its electrical power output is insensitive to day-to-day variations in solar irradiance — or even day-to-night differences!

1.7   Summary of the First Year of Operation of the SEGS-1 Parabolic Trough, Solar Thermal Plant at Dagget, California

A presentation by Eng. Micky Margalith (Luz Industries Israel, Jerusalem, Israel) at the First Sede Boqer Symposium on Solar Electricity Production, 23–24 February, 1986. [English translation from the Hebrew, by David Faiman, authorized by Mr. Margalith.]

1.7.1   *Introduction*

In 1984, the Luz corporation erected the SEGS-1 (Solar Electricity Generating System 1) at Dagget in southern California: a solar power plant with a net rating of 13.8 MW, for a total cost of US$ 62 M of which half the cost was for equipment manufactured in Israel. In 1985 Luz erected a second plant with a rating of 30 MW, which cost US$ 93 M, of which US$ 47 M was for equipment manufactured in Israel. Luz will add another five power plants, each with a 30 MW rating, during the coming three years. Construction of the first of these commenced this year.

1.7.2   Technical details

The solar power plants of Luz Corporation are constructed as modular units, each with a power rating of 30 MW. Erection of each module takes only one year, in such a way that each plant enters service a year after commencement of its construction. This is in contrast to conventional power plants, which are only capable of generating electricity after the number of years it takes to complete their construction.

SEGS-1 includes a net collection area of 71,680 m^2 of parabolic mirrors that track the Sun, and also an energy storage system for increasing the amount of electricity during the hours of peak demand, which do not coincide with the hours of peak solar radiation.

Figure 1.7.1 shows the system, and Fig. 1.7.2 describes, in schematic form, its operation.

The oil, which is heated by the system to a temperature of 307°C, serves to produce saturated steam at 37 bar and 247°C. The latter passes through a gas-fired super-heater where it is heated to 415°C. The super-heated steam is then sent to the turbine.

The system is designed to produce 30,000 MWh of electricity per year, of which 5,000 MWh are produced during the hours of peak demand and 25,000 MWh are produced during other times of the day. Using the energy storage system it is possible to generate [an

Figure 1.7.1:   Aerial photograph of SEGS-1 [Photo Courtesy Luz Corp.]

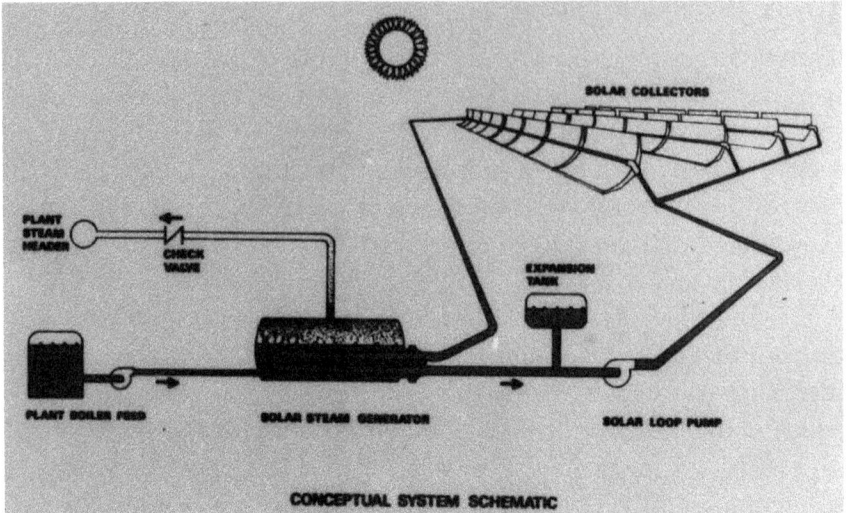

Figure 1.7.2:   Schematic description of SEGS-1

additional] 9,000 MWh during hours of peak demand from energy that was collected during off-peak hours.

The system cost, relative to its output power, is 4.5 US$ per watt of electrical power. The cost of electricity generation (taking account of the cost of system manufacture and construction, land costs, gas, operation and maintenance) is 11.65 US¢/kWh.

The energy storage system added an extra total cost of US$ 8.5 M (construction of the system and maintenance for 30 years). On the other hand, it will increase the expected income by US$ 15.2 M over 30 years, thanks to the additional electricity that will be produced during peak hours.

The storage capacity of the system is 4×10^8 BTU of thermal energy which produces 36.2 MWh of electricity. This output is capable of operating the turbine at full power for nearly 3 hours.

The optical efficiency of the collectors is 75%.

In total, the annual average efficiency of the collector field (output of the field divided by the direct beam solar radiation) is 49%.

The maximum efficiency of the collector field is 66% (for a direct beam radiation of 950 W/m^2 and fluid temperature of 310°C).

The gross turbine efficiency is 31.5%, and 28.5% net.

Overall, the maximum net efficiency of the entire system is 18.8%.

1.7.3   *Performance details*

Figures 1.7.3–1.7.6 show the thermal and electrical production of SEGS-1 compared with the values guaranteed by contract for the first year of operation (during which the guarantee was for [only] 62.5% of the full system output).

In order to raise the system performance a number of improvements are planned for the coming months. Chief among them being:

1) Replacement of a number of heat collection tubes whose vacuum insulation is damaged.
2) Mechanical alignment and adjustment of components (Sun sensors, mirrors, heat collection tubes).
3) Efficiency improvement of the super-heater by adjusting it.

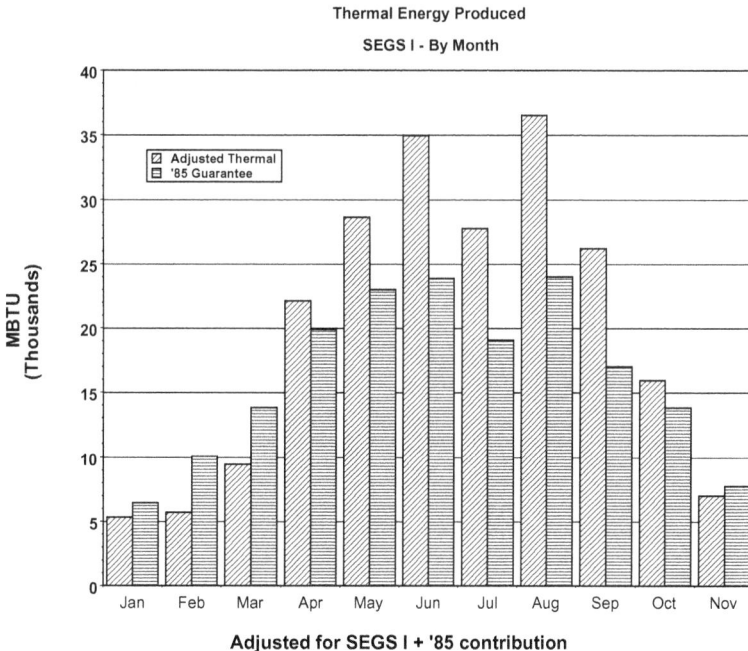

Figure 1.7.3:   Monthly thermal production of SEGS-1

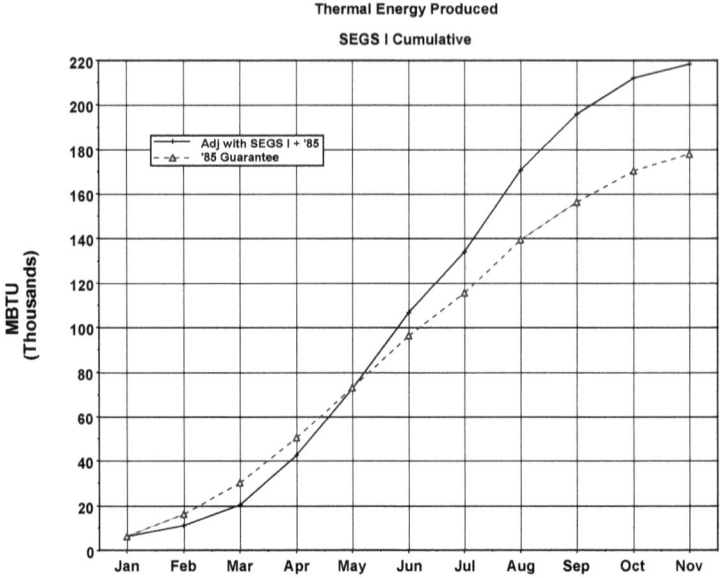

Figure 1.7.4:   Cumulative thermal production of SEGS-1

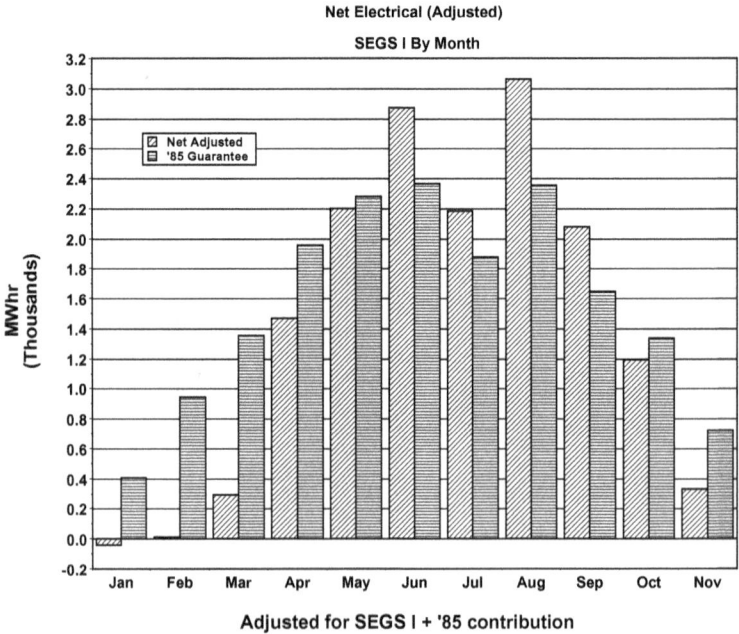

Figure 1.7.5:   Monthly electricity production of SEGS-1

Net Electrical (Adjusted)
SEGS I Cumulative

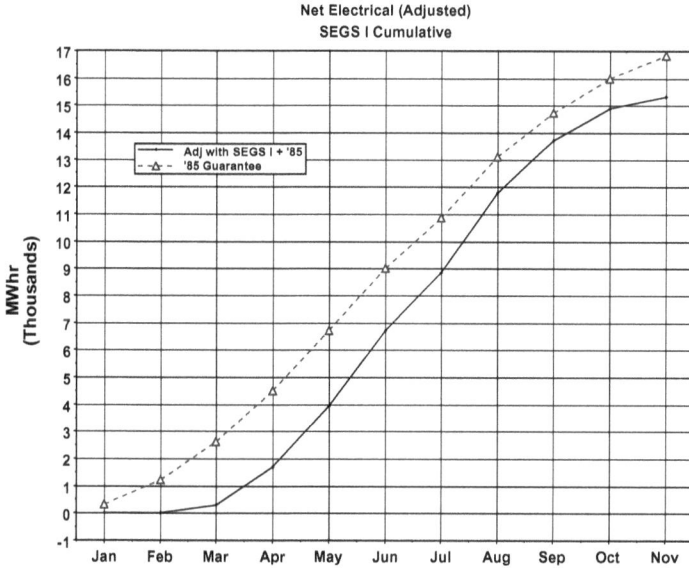

Figure 1.7.6:   Cumulative electricity production of SEGS-1

4) Efficiency improvement of the pumps.
5) Efficiency improvement of the turbine, whose output at present
   is lower than expected (This issue is not the direct responsibility
   of Luz).

1.8   Discussion Following Micky Margalith's Presentation

Q: Why did the amount of electricity generated by SEGS-1 drop
below the guaranteed level for the months of October and November
1985?

Margalith: If you look at the histogram which shows thermal energy
delivery from the collector field you will see that there was no drop
in October and only a statistically insignificant drop for the cloudy
month of November. The poor electrical output was mainly due to
turbine problems for which Luz is not responsible.

By contrast, the start-up of this first-of-a-kind system was plagued
by numerous problems. For example, by April we still only had 520
out of the total 560 collector modules operating.

Q: How much of the total energy output of SEGS-1 is supplied by the gas super-heaters?

Margalith: 18%.

Q: How much of a problem is collector maintenance?

Margalith: The collectors are cleaned approximately every two weeks, at present manually, but in the future this will be automated. Collector efficiency drops by about 8–10% during this two-week period. Cleaning is performed using de-ionized water at a temperature of about 95°C [200°F] and a pressure of about 6 MPa [800–900 psi]. The amount of water involved for each sq. m of collector is 0.64 liter [0.17 gal].

Q: How much does maintenance cost?

Margalith: The entire [annual] maintenance cost — including manpower, replacement parts, materials, etc. — amounts to 1% of the system cost.

Q: How trouble-free is the vacuum system?

Margalith: During the past year we need to replace only 400 out of the total 6720 vacuum tubes.

Q: Can you tell us anything about the economics of SEGS-II?

Margalith: SEGS-II consists of 165,000 sq. m of collectors, i.e. it is more than twice the size of SEGS-1. It is designed to produce 30 MWp and the total installation cost was US$ 93 M. The lower cost per peak watt [US$ 3.1 compared with US$ 4.5 for SEGS 1] is partly due to the use of larger collector modules (5m × 50m compared with 2.5m × 50m for SEGS-1 modules), resulting in fewer motors, etc. We believe that SEGS-II should have a slightly higher efficiency than SEGS-1 but since the system has only been operative for a short period we do not yet have any figures for publication.

Chapter 2

1987

2.1 Editor's Foreword

By the time of the second Sede Boqer symposium, we were fortunate to have attracted two keynote speakers from the world of academia, who were well versed in the respective areas of low-temperature (i.e. T $\leq$ $\approx$100°C) and high-temperature (i.e. T $\geq$ $\approx$300°C) solar-thermal power production. These presentations outline the manner in which system performance efficiency is calculated for the various systems under consideration, and present their relative economic worth at the time those presentations were made. In the case of Professor Rabl, his paper on high-temperature systems has been reconstructed by the present editor, because its author is, sadly, no longer with us. For this purpose, use was made of information from the Professor Rabl's slides, which were reproduced in the conference proceedings, together with a tape recording of his lecture. On the other hand, in the case of Professor Collares Pereira, his presentation on a comparison of various low-temperature solar power-producing systems is the original paper as published in the symposium proceedings.

2.2 High Temperature Solar Thermal Power (Prof. Ari Rabl)

A keynote lecture presented by Dr. Ari Rabl (Center for Energy and Environmental Studies, Princeton University, Princeton, NJ, USA).

2.2.1 *Introduction*

It is always a pleasure to be here at Sede Boqer with my friends and colleagues, especially on such an auspicious occasion as this. I have been following the ups and downs of solar energy and solar power since before the energy crisis of 1973. I have seen, and experienced myself, the hopes, the disillusionment, and the renewed hopes — especially after today's talk by the Luz corporation.

I came to give a review and survey of solar thermal power generated by the high-temperature approach. Preparing this review has been an interesting experience for me because it has given me an opportunity to observe the progress in the various technologies that have been under development for a long time, each racing against the others and proclaiming the superiority of its own approach while denigrating the rivals. Now, quite a bit of time has passed and one can see certain winning design technologies beginning to emerge.

2.2.2 *Solar collectors for electricity generation*

In order to generate electricity from solar heat you must first collect the energy. Several collector types are available, ranked in Table 2.2.1 according to: tracking mode, collector type, temperature range, and suitability for electricity generation.

Table 2.2.1:   Classification of solar collectors

Tracking mode	Collector type	Range	Suitability
Non-tracking	Flat plate Non-evacuated CPC*	<100°C	No
	Evacuated tube + CPC reflector	Approx. 200°C	Perhaps
1-axis tracking	Parabolic trough Linear Fresnel lens	Approx. 350°C	Yes
2-axis tracking	Parabolic dish Fresnel lens Central receiver	Approx. 1000°C	Yes
Other	Fixed reflector + moving receiver	≥250°C	Perhaps

*CPC = Compound Parabolic Concentrator

Without sun-tracking you can only reach about 200°C at reasonable efficiencies. For higher temperatures some form of tracking seems to be necessary. With single-axis tracking you can reach approximately 350°C, and as we heard today from Luz, even the 400°C range has been reached. However, with dual-axis tracking, parabolic dishes, Fresnel lens systems and central receivers can easily be operated at 500–1000°C, and perhaps even 2000°C.

Figure 2.2.1 shows, schematically, a parabolic trough, a parabolic dish and a central receiver with two of its surrounding heliostats.

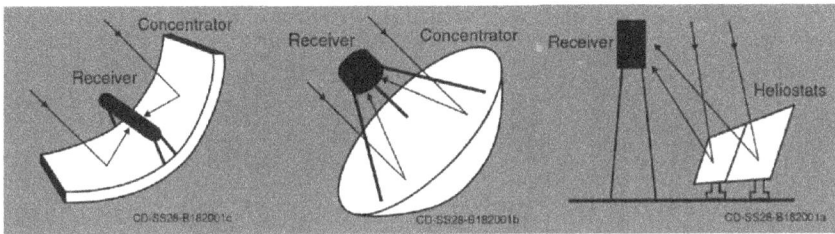

Figure 2.2.1:   Schematic sketches of three types of sun-tracking collector suitable for electricity generation [http://solareis.anl.gov/guide/solar/csp/index.cfm]

The Luz corporation "SEGS-1" and "SEGS-2" are, as we have heard, successful implementations of the 1-axis tracking concept.

A particularly low-cost approximation to the parabolic dish is the system that has been implemented by the LaJet corporation at Warner Springs, CA. It consists of an array of shallow cylindrical drums that are partially evacuated so as to create concave surfaces on the silvered, stretched membranes that cover them. [Fig. 2.2.2].

An operational central receiver system is "Solar 1" at Barstow, CA, which consists of a boiler mounted on top of a tower, 250 meters high. A surrounding field of heliostats each tracks the sun about two axes, redirecting the radiation towards the boiler. [Fig. 2.2.3].

Some other concepts have been pursued, in particular, several types of collector based on fixed reflectors with tracking receivers. The hope here is that the biggest system component, namely, the mirror, is now stationary. When I first came across the concept I was very excited, but it has one fatal flaw: Being fixed, one cannot turn

Figure 2.2.2: The LaJet Corp. 4.9 MW stretched membrane parabolic dish system at Warner Springs, CA [© Peter Menzel/www.menzelphoto.com]

Figure 2.2.3: The "Solar 1" 10 MW central receiver solar thermal plant at Barstow CA [PIX No. 00036, Sandia Natl. Labs., NREL, U.S. Dept. of Energy]

the mirror away from nighttime dew, that brings with it dust, which it glues to the mirror surface. I shall not have anything more to say about such systems, which you may take as a sign of my assessment of their viability.

2.2.3 *Storage options*

Once you have collected the heat energy you can use it right away by converting it directly to electricity, or you can store it for later conversion. Several types of storage are available:

- Oil
- Oil + rocks, glass, etc.
- Molten salt
- None (on the solar side) — Backup with hydro (best!) or fuel.

Several types of oil can store heat at high temperatures, or, to save oil, you can use a tank partially filled with rocks, glass pieces, etc.

It is important to realize that in the near term, the most economical option might be no storage at all, but with backup provided by hydropower, where available, or natural gas or oil. This is because storage comes at a price and you can never extract from storage all the energy you put in. This is particularly true at very high temperatures, i.e. above about 500°C, where storage becomes problematic owing to issues of heat loss and materials stability.

It is also important to realize that storage is only economically viable if the energy, when retrieved from storage, is a cost-effective alternative to fossil fuel. This is not generally the case at nighttime as there is less demand for electricity owing to the fact that fewer people are at work. Consequently, at such times the price of fossil-fueled electricity is low. The key to storage strategy would thus seem to follow the apocryphal words of the banker, J.P. Morgan, whose motto is said to have been "buy cheap and sell dear". So, in practice, one only needs a few hours of storage: a situation that will continue until such time as stored solar heat might become more economical than the nighttime use of fossil fuel.

2.2.4 *Conversion of solar heat to electricity*

In order to convert the collected heat to electricity, one must employ a heat engine that typically operates using a Rankine, Sterling, Brayton, etc. cycle. Moreover, one has heat transfer across heat exchangers to contend with, which lowers the available temperature potential, thereby yielding a conversion efficiency that is significantly below the Carnot limit. Nonetheless, as a general first guide it is instructive to think about the Carnot efficiency.

Suppose we set the low temperature sink at 20°C. This is the number I usually use for calculations in New Jersey. Perhaps, here at Sede Boqer, it would be higher. If the high temperature source is at 90°C — as it might be for a good flat-plate solar collector, then the Carnot efficiency is around 20%. In practice you could expect to get up to about two thirds of the Carnot efficiency. The reduction is actually greater at lower temperatures, so you really need to have a heat source that can supply heat at around 500°C. Table 2.2.2 illustrates these points.

Table 2.2.2: Carnot efficiencies for high temperature heat sources and 20°C heat sink

T(high) [°C]	90	200	500	1000
Carnot Efficiency [%]	19	38	62	77

Incidentally, 500°C is approximately the temperature at which conventional central power stations operate using steam turbines, and they can achieve about 40% efficiency in practice. Such plants are fired by coal, oil or nuclear heat. Their turbines operate in a cycle — known as the Rankine cycle, after the person who first analyzed it. The heat first vaporizes a liquid in a boiler. The subsequent expansion of the vapor performs work on the turbine. The condensation of the vapor back to a liquid, in a cooling tower, and its return to the boiler completes the cycle.

Another very useful cycle is the Brayton cycle, employed in gas turbines, and also in aircraft jet engines. Incidentally, the "gas" in gas turbines refers to the working fluid: not the fuel. In a gas turbine, the gas is first compressed and then allowed to expand through a turbine. It is then recompressed, without any change of phase. That is to say, it remains a gas throughout the cycle.

Yet another thermodynamic cycle that has been proposed and is under development is the so-called Stirling cycle, which is a reciprocating external combustion engine, i.e. it has an external heat source.

There are also a few other more exotic approaches, such as magnetohydrodynamics, which might perhaps be promising for the future.

2.2.5 *System considerations*

Having now examined the components let us think about the various ways they can be combined to create a system. Not all combinations will be practical owing to component cost considerations, and also to the match between the performance of the various engines and the pattern of electricity consumption. There are also economies of scale, efficiencies of scale, questions of reliability, etc.

2.2.5.1 *Carnot restrictions on collector performance*

The Carnot limit implies the need for either high temperatures (i.e., at least 500°C) or very low-cost collectors (e.g. solar ponds, or ocean energy). Regarding cost, you need to keep in mind that conventional power plants have a fuel-to-electricity efficiency of 30–40%, which leads to an electricity cost of approximately 3× the fuel cost. Therefore, if a solar collector is to be competitive with fuel as a heat source, its heat-to-electricity efficiency must be at least 30%. This is not possible with flat-plate or even CPC-mounted evacuated tubes — unless their cost can be appropriately reduced.

2.2.5.2 *Efficiencies and economies of scale*

2.2.5.2.1 Collector size

There are obvious economies of scale associated with large solar collectors: fewer moving parts, etc. However, they are also subjected to greater wind forces and weight stresses. At present, the optimal length of such units is thought to be in the 5–15 m range for dishes and heliostats, i.e. collector areas in the range 50–150 m^2. However, this optimal length is based on economic considerations, and increases with production volume.

2.2.5.2.2   Turbine size

The steam and gas turbines that are used in central power plants are most efficient when they are built in large scale — at least megawatts. In fact, the increase of efficiency with scale for steam turbines has led to the development of very large power plants — as large as 500–1000 MW. Large gas turbines also exhibit increased efficiency, but they are mostly used for producing power during peak periods and short periods of peak consumption.

Incidentally, one should bear in mind that a turbine capable of producing 1 MW of electricity requires a solar collection area of roughly 70 m × 70 m. This is a bit too large to be placed on a single tracker. This consideration tells you that such turbines are only suitable for large central systems where the heat is collected to a single central point.

Stirling engines, on the other hand, can achieve high efficiencies on a much smaller scale. They are therefore capable of being mounted on a parabolic dish, a collector type that is well suited for producing the 800°C temperatures that enable such engines to operate at high efficiency. At the present time Stirling engine technology is not quite commercial but a lot of work is going on to develop them. Percentage efficiencies in the low 30s have been reported but it is hoped to reach 35–40%. Such engines would have not only solar applications but also various mobile uses such as the replacement of conventional automobile engines.

2.2.5.3   *Collector thermal considerations*

2.2.5.3.1   Collectible heat

Of the various collector technologies, the parabolic dish collects the most solar radiation per unit collection area because it is always normal to the sun.

The central receiver and the parabolic trough both collect approximately 2/3 the amount of radiation as the dish per unit collection area. The precise ratio is dependent of geographic latitude and spacing, which give rise to mutual shading, blocking and cosine effects. However, the mirrors needed for these two latter collector types are simpler and hence of lower cost than for a parabolic dish.

In fact, the mirrors can even be flat for the central receiver type. This is a significant cost saving factor compared to dish and trough collectors where the mirrors need to be accurately shaped. Another cost advantage of the central receiver is that the collected energy is transported optically to a central point, thus reducing the cost of pipes and their associated losses. The only pipe is that which transports the heat from the central boiler to the ground.

Piping costs and their associated heat losses are moderate for trough technology because most of the piping is the receiver tube.

Dish technology is the most problematic, both because of the extensive piping needed to transport the hot fluid to a central turbine, and to the need for increased insulation to minimize heat losses from the higher receiver temperatures that dishes can reach (unless their working temperature is restricted to around 300°C). It is a concern for such losses that makes Stirling engines such potentially attractive receivers for mounting on parabolic dishes. With such a field, it would be electricity rather than heat that would flow from each dish.

2.2.6 *Promising options*

After extensive analyses that fill hundreds of reports, the most promising combinations at present seem to be, for the near term, two technologies that are now commercial: namely, the parabolic trough plus steam turbine systems that have been demonstrated successfully by the Luz corporation; and the LaJet parabolic dish system, which has been modified from its original design in order to produce steam at about 300°C.

It is important to note that in the near term, parabolic troughs (and dishes that are restricted to operate at similarly low temperatures) provide a very efficient match to the temperature needed to evaporate the water that feeds conventional steam power plants. Such plants expend about 80% of their heat energy for this purpose, i.e. at temperatures below 350°C, and the rest for superheating the steam to 500°C. Thus a parabolic trough system can efficiently provide the lion's share of the energy for a conventional steam turbine. The remaining energy can then be provided by a gas superheater in series with the solar collectors.

Such a system has the additional advantage of being able to generate power when the sun does not shine, or from storage, thus gaining capacity credit for electricity generated during the entire peak period in the late afternoon and early evening hours.

This is perhaps a good place to mention something I learned from speaking to the Luz people at this symposium. At first, I was surprised to learn that they operate their solar collectors at 400°C, as they are obviously less efficient at such a high temperature. However, it turns out that the overall economics is dictated by the arbitrarily small amount of gas per year that they are contractually allowed to use. Careful analysis of the regulatory requirements and the available tax breaks, then prompts Luz to operate the solar collectors at less than their maximum efficiency.

Another promising technology is the central receiver coupled to a steam turbine with, perhaps, a few hours of thermal storage. I should emphasize that this technology is not yet commercial, but holds out much promise for the future. One of the reasons is that the high temperatures that can be reached may obviate the need for a fossil-fueled superheater.

A second reason is that the high temperatures that can be reached by central receiver systems could drive a gas turbine directly, as the latter operates at typically 1000°C. Moreover, no cooling tower would be necessary because, as with jet engines, the atmosphere could provide a low temperature heat sink for the exhaust heat at 500°C.

Yet a fourth promising technology is the Stirling engine, as such a unit could operate at the focus of a parabolic dish. But here the technology requires much further development, considerably higher temperatures are involved, typically 2000°C, which lead to many material issues.

2.2.7 Current research in the USA funded by the US DOE

2.2.7.1 Reflectors

The US Department of Energy (DOE) is funding much research into low-cost reflective materials. Silver is the obvious choice for the reflector because of its high reflectivity. But it needs to be protected. Glass provides such protection when silvered on its rear surface, but we need to find lower-cost alternatives to glass.

At present the emphasis for heliostats (the single most costly component of a central receiver system) seems to be on stretched steel membranes, with a silver upper coating and some form of transparent or semi-transparent polymer coating for protection. These collectors would be partially evacuated so as to provide a small amount of curvature for concentrating the reflected radiation.

In Fig. 2.2.4 I have outlined the DOE's roadmap to the year 1990 for heliostat development along with some cost projections. These projections start at today's cost of around US\$ $900/m^2$, down through perhaps US\$ $80/m^2$ for metal-on-glass units, and then down to perhaps half that figure for stretched membrane heliostats.

2.2.7.2 Receivers

Here the emphasis is on direct absorption of radiation by the heat transfer fluid. This results in increased efficiency as it removes the need for a heat exchanger and its associated losses. Examples are the use of molten salt, or suspended small particles, such as carbon, in an air stream.

2.2.7.3 Second-stage concentrators

Work is also being carried out on second-stage concentrators, which effectively increase the concentration ratio and decrease the heat loss.

2.2.7.4 Ceramic heat-exchangers

Ceramic heat-exchangers are also under development, in order to enable efficient use to be made of the extremely high temperatures that are potentially available from central receivers. At present,

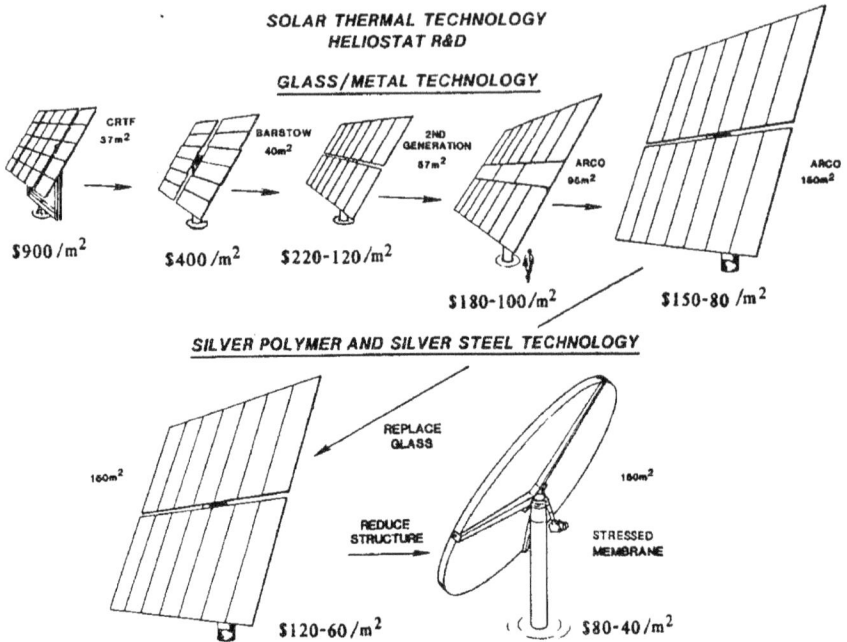

Figure 2.2.4:   USDOE projections for heliostat technology development [Reproduced from: "Five Year Research and Development Plan 1986–1990" U.S. Department of Energy, DOE/CE-0160, Sept. 1986]

materials considerations dictate a maximum working temperature of about 500°C from the latter. But appropriate heat exchangers could extend the useful range of storage, by allowing the storage to be charged at much higher temperatures. This would result in higher exit temperatures toward the end of the period, when storage is almost exhausted.

2.2.7.5 *Engines*

Research on engine development is concentrating on Stirling engines. I have already emphasized their potential importance for electric power production, both as receivers for individual parabolic dishes, and as a high-temperature alternative to a steam turbine for central receivers. But much work is still necessary to improve their efficiency and reliability.

2.2.7.6 *Processes*

Although this is not strictly power production, the use of highly concentrated solar flux, as produced by parabolic dishes, could replace the use of electricity for several industrial processes such as the production of hydrogen, the gasification of biomass (and of toxic wastes). Perhaps in the future it might also be possible to produce metals and other pure minerals from ores.

2.2.8 *Existing solar thermal power plants*

Let us now take a look at the solar-thermal power plants that have actually been built. I have collected them in Tables 2.2.3, 2.2.4 and 2.2.5, sorted according to technology type. The most interesting plants are the large ones for which costs are given in units of US dollars per peak electrical watt [US$/We,pk]. Such cost figures are only approximate indications of true costs because of differing marketing strategies, tax breaks, and local regulatory factors. Also, you must take into consideration the fact that the costs of this first generation of solar plants include the research and development costs. For example, it is estimated that a repeat of the Barstow plant, which in Table 2.2.3 is given as 14 US$/We,pk might be as low as 3 US$/We,pk. However, there are no plans at present to replicate Solar 1.

Table 2.2.3: Central Receiver plants

Plant	Year complete	Aperture [m^2]	Peak power [MWe]	Cost [US$/We,pk]
"Solar 1" Barstow, CA, USA	1982	73,000	10.0	14*
"Themis" Targassone, France	1983	11,000	2.5	12
"CESA 1" Almeria, Spain	1983	12,000	1.0	15
"CRT" Nio, Japan	1981	13,000	1.0	26
"Eurelios", Sicily, Italy	1981	6,000	0.6	12
"CRS-1" Almeria, Spain	1981	4,000	0.5	30

*Might be reducible to 3 US$/We,pk in the future.

Table 2.2.4:   Parabolic Dish plants

Plant	Year complete	Aperture [m²]	Peak power [MWe]	Cost [US$/We,pk]
"Solarplant-1" (LaJet) Warner Springs, CA, USA	1984	30,000	5.0	3.8
"STEP" Shenandoah, GA, USA	1983	4,000	0.4	?
"Sulaibiya", Kuwait	1981	1,000	0.1	?

Table 2.2.5:   Trough plants

Plant	Year complete	Aperture [m²]	Peak power [MWe]	Cost [US$/We,pk]
"SEGS 1", (Luz) Dagget, CA, USA	1984	72,000	14.0	4.2
"SEGS 2", Dagget, CA, USA	1985	165,000	30.0	?
"SEGS 3", Dagget, CA, USA "SEGS 4, 5, 6, ... " (?)	1986	200,000	30.0	3.3
"Vignola" Corsica, France	1981	1,000	0.1	?
IEA "DCS" Almeria, Spain	1981	5,000	0.5	?
"PPM" Nio, Japan	1981	11,000	1.0	?
"Coolidge" AZ, USA	1980	2,000	0.2	26
"Meekatharra" Australia	1982	1,000	0.1	36

More interesting are the stated costs of the 5 MW LaJet dish plant, and the Luz trough plants. For the latter, we can already see a cost reduction from 4.2 US$/We,pk for SEGS 1 to 3.3 US$/We,pk for SEGS 3, and we are told that they will come down to 2.5 US$/We,pk for the next Luz plants. One of the impressive features of the Luz plants is that not a penny of government financing went into their development or production. The SEGS plants that were built for the

Southern California Edison electric company were all financed by private investment.

2.2.9 *Economics*

There are many ways of performing an economic analysis, depending on the accuracy you require and the factors you consider as being important. However, let me give you a simple rule of thumb which I find useful for assessing the value of solar plants.

You start with the so-called Capacity Factor, which is the ratio of the average plant output to its peak power output:

$$CF = kWe, av/kWe, pk.$$

The Capacity Factor may either be given by the manufacturers, or can be estimated on general grounds. The annual output for a 1 kWe,pk plant [in kWh] is then:

$$Q = CF \times 8760 \, hrs/yr \times 1 \, kWe, pk.$$

The simple rule of thumb, which I refer to as "the 10% rule", is to divide Q by the annual cost of the electricity, which is estimated by taking the plant cost and dividing it by 10.

Why 10%? As it happens, this closely corresponds to a loan at 8% interest for 20 years. This is a realistic rate, which would include interest, repayment of the loan, plus a small addition for operation and maintenance. Of course, it assumes that the plant will continue to operate for 20 years at its rated performance and without breakdowns.

As an example, take the LaJet plant. For this kind of plant, CF should be approximately 0.25, giving an annual electrical output per kWe,pk of 2190 kWh. 10% of the plant cost is 380 US$/kWe,pk. Therefore, the expected electricity cost is 17 US¢/kWh.

2.2.9.1 *The value of solar electricity*

The cost of solar-generated electricity must, of course, be compared with the cheapest conventional alternative: otherwise, nobody would invest in the solar power system. In the United States, off-peak electricity between the hours of about 10 pm to 7 am is generated by coal or nuclear fuel, at a cost of roughly 4 US¢/kWh. This figure is based on roughly 2–2.5 US¢/kWh for coal, and 1.5–2 US¢/kWh for capital costs. On the other hand, peak electricity, generated by gas turbines and more expensive fuel, costs about 8 US¢/kWh on hot summer afternoons. So the solar electricity generated by the LaJet plant at such times is still too expensive by a factor of approximately two. Consequently, there is a need for some form of subsidy at the present stage of development.

However, costs are expected to come down. For example, a replica of the Barstow central receiver plant would generate electricity for a cost of about 12 US¢/kWh, and US Department of Energy projections suggest that the cost would fall to about 5 US¢/kWh by the year 2000.

2.2.9.2   *The learning curve*

I want to end by saying something about an interesting empirical regularity that has emerged from studies of industrial development across many sectors of industry, e.g. airplane production, radios, refrigerators, etc. It reflects the obvious fact that the more you build, the more you learn how to do it more efficiently. It is called the "learning curve". The observation is that there is a reduction in unit cost by a factor of approximately 0.7 to 0.9 for each doubling of cumulative total production.

Let me take the case of Luz as an example, because there are clear indications for the start of a replication process. If we start with SEGS 1 at a cost of 4.2 US$/We,pk, and take a learning curve reduction factor of 0.9, the successive amounts of cumulative production and corresponding expected costs are shown in Table 2.2.6.

Table 2.2.6:   Learning curve projections for SEGS plants

Cumulative Production [MW]	14	28	56	112	224	448	896
Expected Cost [US$/We,pk]	4.20	3.78	3.40	3.06	2.76	2.48	2.23

Interestingly, SEGS 3, at 3.3 US$/We,pk, for a cumulative production of 74 MW, fits nicely within this range. However, one must realize that learning curves in practice can be very jagged rather than smooth, and they can only be observed in retrospect. They cannot form a basis for prediction.

2.3 Discussion Following Ari Rabl's Presentation

Faiman: Why do you address the economics of solar-produced power from the simple over-simplistic viewpoint of fuel saving?

Rabl: It is obviously not the whole story, as you clearly observe. Capacity credit is, however, an extremely complex issue and it does not clarify the economics. It is nevertheless, meaningful to compute the cost of producing solar electricity as a separate issue from how much a utility is prepared to pay for the product. As we have seen, today's solar-produced electricity appears to cost about twice as much as when it is produced by conventional means. This is an encouraging development.

Nir: I have two comments: First, the use of a learning curve for predictive purposes is fraught with uncertainties owing to the possibility that some new discovery may dramatically alter the cost trend. One has only to look at the effect environmental considerations have recently had in the energy sector.

Secondly, I am disturbed that analyses such as yours always seem to compare the projected cost of some alternative energy source with the present (rather than the projected) cost of conventionally-produced energy. I am aware that electricity production seems to be a rather mature technology, but it should not be ruled out that advances in the technology of storage may radically lower the cost of conventional electricity.

Weiner: One developing technology that appears promising is that of the gas turbine.

Rabl: I agree, although I do not know of any plans to combine solar and gas turbine technologies.

Charatz: Luz is looking into that possibility.

Roy: To what extent does the cost of achieving optical precision influence the relative economics of the various kinds of concentrators?

Rabl: Judging by Luz's commercial success, the optical precision of today's parabolic troughs would appear to be adequate in the 300°C range of temperatures. For the 500°C range, the present generation of parabolic dishes also has adequate optical accuracy. For higher temperatures it appears that a second-stage concentrator will be necessary to compensate for aberrations in the primary reflector. Whether this added complexity will be economically justified — only time can tell.

As to the question of an economically optimal temperature for solar-thermal power production: today the answer seems to reside with Luz. Tomorrow, it might lie with central receivers — but if they are not built, that too will be part of the answer.

2.4   Low Temperature (T < 100°C) Solar Thermal Electricity (Prof. Manuel Collares Pereira)

A keynote lecture presented by Dr. Manuel Collares Pereira (LNETI, Lisbon, Portugal).

2.4.1   *Introduction*

The objective of this talk is to review existing solar energy technologies for the production of low temperature heat (T < 100°C) in terms of their potential to generate electricity.

These technologies will be compared on equal grounds, namely in terms of annual costs expressed in US\$/kWh, assuming 20 years of operation. In order to do this comparison we first estimate how much energy they can produce on a yearly basis and then we make educated guesses of their present costs (as possible costs, assuming acceptable projections from today's costs in the case of those technologies that are not usually used for the production of electricity).

Rather than comparing these costs with conventional energy costs we decided to compare them with present-day energy costs from stationary photovoltaic (PV) systems, in order to place in perspective the interest of pursuing efforts in low temperature heat for electricity. The results obtained show PV as already having a lower cost than other technologies presented, and since the costs for the thermal technologies are much closer to their limiting values, this likely means that in the future, low-temperature solar heat should be more used directly as such, i.e. rather than for the production of electricity.

The talk is organized as follows:

- A brief review of how to convert low-temperature heat to electricity, the Rankine Cycle;
- A brief description of how to calculate yearly average system performance;
- A brief description of the various technologies considered, namely ponds (salt gradient and others), solar collectors [as a specific example: a non-evacuated compound parabolic concentrator (CPC) plus storage], and photovoltaic systems;
- A calculation of yearly system performance in each case;
- An estimate of each system's cost and a comparison of all systems in terms of US$/kWh produced;
- A discussion of the results.

2.4.2   *Conversion of low-temperature heat into electricity via a Rankine cycle*

The appropriate thermodynamic cycle for conversion of low-temperature heat is a Rankine cycle [1].

It is assumed in the following calculations that the temperature difference between the hot and cold sources is $\Delta T = 55\,\text{K}$, with typical yearly average values of 85°C at the boiler, and 30°C at the condenser. Carnot efficiency in this case is around 0.15, and a *practical limit* for the Rankine cycle efficiency is roughly half of this figure, i.e. $\eta_R = 0.08$.

In the literature and for the temperatures conditions stated above the typically reported η_R is 0.05. Recently, $\eta_R = 0.06$ was reported [2], which is a good (albeit high) value.

2.4.3 *Yearly average energy performance of solar systems*

The instantaneous efficiency of solar collectors can be described in terms of the Hottel-Whillier–Bliss equation, namely:

$$\eta = F_R \eta_O - F_R U_L \times (T_{in} - T_{amb})/I$$

where the T_{in} is inlet temperature to the collector, T_{amb} is the ambient temperature, I is the instantaneous radiation on unit area of the collector plane, η_O is the optical efficiency, U_L is the collector heat loss factor and F_R is a factor which takes care of the fact that the collector's performance is being described in terms of T_{in} and not in terms of the receiver temperature.

The average energy delivered by a collector operating in a system at a constant inlet temperature can be calculated by

$$Q_{year}(T_{in}) = F_R \eta_O \times H_{coll} \times \phi(X_{in}).$$

As written, Q_{year}, which is in units of $MJ/m^2/day$, must be multiplied by 365 for the total yearly energy. H_{coll} is the average radiation available to the collector per unit area of entrance aperture, η_O is the average optical efficiency (including incidence angle modifier), and $\phi(X_{in})$ is the utilizability, a function with a value <1 accounting for the fact that radiation below a certain threshold X_{in} is not usable, because it does not overcome the collector's heat losses.

$$X_{in} = U_L \times (T_{in} - T_{amb})/\eta_O.$$

In all that follows the radiation data used is from Bet Dagan, Israel (Latitude $= 32°$).

2.4.4 *Low temperature solar heat technologies*

There are two general types of technology: those where collector and storage are combined, e.g. solar ponds; and those where the collectors are separated from storage, which will be referred to as collectors plus storage.

2.4.4.1 *Solar ponds*

Of all solar ponds, salt gradient solar ponds [3] are the prime candidates for collecting and storing solar energy at temperatures suitable for its conversion to electricity via the Rankine cycle described in II, since they can be constructed and operated at costs which are among the lowest possible for solar collectors. The Rankine cycle uses the hot water from the lower convective zone of the pond as the heat source, and the cooler surface water as the heat sink.

Other ponds have been proposed, such as shallow ponds, or ponds covered by opaque insulating materials, but their thermal performance is inferior to that of the salt gradient variety. In the analysis below we shall consider the salt gradient pond in great detail and the shallow pond as an alternative example.

2.4.4.2 *Solar collectors: Non-evacuated CPC plus storage*

The other main type of system to be considered is one in which solar energy is collected in active solar collectors by heating a fluid which will be stored separately, until used as the heat source in a Rankine cycle.

Because the collectors must heat a fluid to at least 85°C with reasonable efficiency and cost, we choose to use in this study non-evacuated CPC collectors with a concentration ratio of 1.2× [4]. They are able to perform with efficiencies above 0.5 up to temperatures around 100°C and can be manufactured at the cost of a good, selectively coated, single glazed, flat-plate collector.

2.4.4.3 *Photovoltaic systems*

For comparison we consider a *PV* system with an array of modules of the flat-plate, monocrystalline silicon type, mounted, equator-facing, at a tilt angle equal to the latitude of the site. The array is connected to an inverter and then to the grid. However, and as in the other cases, we will not consider the transformer as part of the system.

2.4.5   *Yearly system performance*

2.4.5.1   *Salt gradient solar pond*

Following Kooi [5] we can now estimate the best possible performance of a salt gradient solar pond of area $1\,\text{km}^2$ for the conditions stated above. We will assume that the ΔT of $55\,\text{K}$ is maintained all year round (This really means for instance, that in winter time, the temperature of the lower convective zone can drop to $70°\text{C}$, and ambient surface water temperature will be around $15°\text{C}$). The Rankine cycle will only operate whenever this ΔT is achieved, which allows us to think in terms of a radiation threshold and to use the utilizability concept. However, nighttime losses must be explicitly deducted from the value of Q_{year} found in the manner described above.

At normal incidence and for pure water [15] we may have $F_R\eta_O = 0.35$ (note that this is a high value since typically in clear and clean ponds $0.25 < F_R\eta_O < 0.30$). $F_R U_L$ can be assumed to be at best $0.64\,\text{W/m}^2/\text{K}$. [We are thinking of a pond with a gradient zone of $1\,\text{m}$ + upper convective zone of $0.5\,\text{m}$; this is again an optimistic estimate since typically, $1 < F_R U_L < 2\,\text{W/m}^2/\text{K}$.]

Radiation does not come at normal incidence but there is an effective angle, β_{eff} (energy weighted) at which, we can assume, all incident radiation reaches the water's surface: $\beta_{\text{eff}} = 52°$ for the case of Bet Dagan. There is a reduction of 5% resulting from it. Also, the path length of the refracted radiation in the water is accordingly increased, resulting in further absorption: an approximate 4% effect.

We then have [where $\langle x \rangle$ represents the average value of x]

$$\langle F_R\eta_O \rangle = F_R\eta_O \times 0.95 \times 0.96.$$

From the radiation data we have

$$\phi(X_{\text{in}} = 92\,\text{W}) = 0.65$$

and

$$\langle H_{\text{coll}} \rangle = 19.2\,\text{MJ/m}^2/\text{day}$$

and therefore

$$\langle Q \rangle_{\text{year}} = 3.9\,\text{MJ/m}^2/\text{day}$$

from which, nighttime losses must still be deducted, i.e. $Q_{loss} = 1.5\,MJ/m^2/day$.

Assuming that $\eta_R = 0.08$, we obtain an upper limit for the amount of electricity produced of

$$\langle Q \rangle_{elect} = 0.08 \times 2.4\,MJ/m^2/day = 0.19\,MJ/m^2/day.$$

However, the available Rankine cycles have, as stated, lower efficiencies and therefore a more realistic value for Q_{elect} is

$$\langle Q \rangle_{elect} = 0.05 \times 2.4\,MJ/m^2/day = 0.12\,MJ/m^2/day.$$

This last figure is consistent with measurements presented by B. Doron at the first symposium in this series, for the $200{,}000\,m^2$ pond at Beit HaAravah [2].

However, to operate the cycle it is necessary to consume a fraction of the electricity produced. We shall assume that it is necessary to spend 30% [3] of the electricity produced in order to power the cycle. This seems to be an optimistic assumption since the reported consumption at Beit HaAravah was much closer to 50% [2].

We have then

$$\langle Q_{elect} \rangle_{max,net} = 0.13\,MJ/m^2/day.$$

$$\langle Q_{elect} \rangle_{present,net} = 0.08\,MJ/m^2/day.$$

2.4.5.2 *Shallow pond*

Based on measurements done at Lawrence Livermore Lab, USA, we have

$$F_R\eta_O = 0.74, \quad F_R U_L = 7.4\,W/m^2/K.$$

With $\phi(X) = 0.08$ we have $\langle Q \rangle_{year} = 0.67\,MJ/m^2/day$ and $\langle Q_{elect} \rangle_{max,net} = 0.035\,MJ/m^2/day$.

2.4.5.3 *Non-evacuated 1.2X CPC*

In this case the collector parameters used correspond to a particular, simple-covered, collector manufactured in Portugal [7]:

$$F_R\eta_O = 0.74, \quad F_R U_L = 2.84\,W/m^2/K$$

Assuming that the collectors are operated at tilt = latitude, we have

$$\langle F_R \eta_O \rangle \times \langle H_{coll} \rangle = 12.8 \, \text{MJ/m}^2.$$

$$\phi(X) = 0.58.$$

$$\langle Q \rangle_{year} = 7.4 \, \text{MJ/m}^2.$$

From this we derive

$$\langle Q_{elect} \rangle_{max} = 0.08 \times 7.4 = 0.6 \, \text{MJ/m}^2$$

and

$$\langle Q_{elect} \rangle_{present} = 0.05 \times 7.4 = 0.4 \, \text{MJ/m}^2.$$

In this case operational experience with collector fields of the kind considered here, plus the power consumption for the operation of the cycle proper, leads us to estimate a parasitic power consumption of 20%, from where we obtain

$$\langle Q_{elect} \rangle_{max,net} = 0.5 \, \text{MJ/m}^2,$$

$$\langle Q_{elect} \rangle_{present,net} = 0.3 \, \text{MJ/m}^2$$

in the two cases above.

2.4.5.4 *Photovoltaic system*

We assume an efficiency for the PV system of 0.1. $\langle H_{coll} \rangle$ at tilt = latitude and zero azimuth is $20 \, \text{MJ/m}^2/\text{day}$ resulting in $\langle Q \rangle_{year} = 2 \, \text{MJ/m}^2/\text{day}$ [8].

We can collect all the results above into Table 2.4.1 where we show the average net electrical energy produced per day per m², by each system both in the "maximum" and what we have denoted the "present" situation.

2.4.6 *Estimate of system costs*

In what follows, we shall assume that all land is free, and that salt transport is free (assumptions which favor the low temperature collectors, especially the solar ponds).

Table 2.4.1: Annual expected electrical output per m^2 of various collector types

MJ/m^2/day	Salt gradient solar pond	Shallow pond	CPC + storage	PV
Maximum	0.13	0.035	0.5	—
Present	0.08	—	0.3	2

We first estimate each system's capital costs and then compare systems, calculating annual costs with the Rabl "10% rule", i.e., the total annual costs for interest, repayment of loan as well as Operation and Maintenance are 10% of the initial investment [9], assuming that the systems function reliably for at least 20 years.

Common to the low temperature systems is the cost of the Rankine generator + associated equipment. For the sake of the comparison we will take a 1 Km^2 area with which it is proposed to associate a 5MW Rankine engine [3] with a present total cost of US\$1500/kWp (lowest estimates are of US\$1000/kWp but we will retain the higher one).

2.4.6.1 *Salt gradient solar pond*

From a thorough review of pond related papers (for instance [10, 11]) and from my own experience in the subject (for instance [12]) and from extensive published material on the construction of water basins, my lowest estimates of pond costs in US\$ per m^2 are as follows (Table 2.4.2):

Table 2.4.2: Expected costs in US\$/m^2 associated with lined and unlined solar ponds

Item	Lined pond	Unlined pond
System components	13	9
Water + chemicals (20 years)	6	
Fence	2	
Excavation	0.7	
Total	21.7	17.7

These figures should be compared with the typical minimal costs considered all over the world [10, 12] of US\$50–70/m^2, which, however, include the cost of salt, salt transport, and a leak-proof construction given potential pollution problems in areas where contamination of the water table from salt could not be tolerated.

From Table 2.4.2, retaining the lowest number and adding the Rankine engine cost, we obtain US\$24.8/m^2.

2.4.6.2   *1.2X CPC + storage*

We assume that for a 5 MW system we would be able to have a US\$130/m^2 installed system (US\$120/m^2 for collector frame connecting pipes and control [7] and US\$10/m^2 for storage). To this figure we must add US\$37.5/m^2 for the Rankine engine as before. [This number was obtained taking into account that the CPC system produces 5× more energy on a per square meter basis than the solar pond.] Therefore, the total capital cost is US\$167.5/m^2.

2.4.6.3   *Photovoltaics*

We assume [8] a US\$5/W_p for the panels and US\$0.5/W_p for frame, inverter, wiring, power tracking, etc., for a total cost of US\$5.5/W_p which in our case, and given that peak power per m^2 is 100 W, means a total cost of US\$550/m^2.

Applying the 10% rule referred above, using Table 2.4.1 and the costs described in this section we list in Table 2.4.3 the annual costs in US\$/kWh of electricity produced in each case in both "maximum" and "present" situations.

Table 2.4.3:   Estimated costs in US\$/kWh of electricity from various system types

Cost$_{annual}$	Salt gradient pond	Shallow pond	1.2X CPC	PV
"maximum"	0.19	—	0.33	
"present"	0.30	—	0.55	0.27

2.4.7   *Discussion of results*

Table 2.4.3 shows that of the thermal options retained it is clear that salt gradient solar ponds are the best. However, PV is already cheaper and its cost is likely to go much further down than the limiting lowest possible cost for ponds (0.19 US$/kWh) estimated here. This seems to point towards the prediction that, at least in the medium term, electricity from low-temperature heat sources will not be of interest.

A word should be said about ponds and the fact that they can store energy from one day to the next and indeed even for several days, thereby having a potential usage as "stand alone" devices, a situation that was not considered here. They will not, however, deliver more energy in that mode (on the contrary, there will be more losses and possibly less gains from an increase of the threshold). The problem with salt ponds is that at these costs/kWh they are really site limited to those places where there is free flat land, free salt water, and to low latitudes (Lat $<35°$ because of the angle the sun makes with the water surface during each day and in the course of the year), conditions that are only fulfilled by less than a handful of spots on Earth.

Photovoltaics clearly does not have these limitations and can be installed on existing structures, rough terrain, etc. Also, while comparing ponds and PV in grid connected situations (as done above) it can be argued that ponds can produce all of their daily electricity a few hours later in the day or night, therefore having a higher value because of their natural adaptation to a "late" peak demand. However, including a couple of hours of battery storage in the analysis above for the PV system would not significantly increase the costs and would have the same effect. Besides, in sunny locations, there seems to be a match between peak demand and peak power availability from the sun, thereby reducing the importance of the above argument.

Therefore, it seems that low temperature heat with ponds or collectors will have a much more widespread use directly for heating applications like: domestic and industrial-process water heating, agricultural applications (greenhouse heating, water purification, or solar drying, for instance) and, even, in the case of ponds, applications like the direct industrial extraction of chemicals from mineral ore [13, 14].

2.4.8   Note added in proof (i.e. author's original proofs for the Symposium Proceedings)

At the Symposium, Paz Company claimed that their own selling costs for a tracking PV system installed, grid connected (transformer excluded) are 5.5 US$/Wp. In their case, Q_{year} should be higher than that of the stationary PV system considered here, reducing the figure in Table 2.2.3 for PV costs by a factor between 1 and 2.

2.4.9   Acknowledgements

I wish to thank Dr. John Hull for providing much of the information used to find out about pond costs, and Dr. Ari Rabl, Prof. Jeff Gordon and Prof. Yair Zarmi for the discussions concerning the assumptions and results presented in this lecture.

2.4.10   References

[1] "Application of solar technology to today's energy needs" Office of Technology Assessment, Congress of the United States, Vol. 1, 1978.

[2] B. Doron, "Proceedings of the First Sede Boqer Workshop on Solar Electricity Production", 23th Feb. 1986

[3] H. Tabor, "Solar Ponds", Second International Symposium on Non-Conventional Energy — Trieste, 1981. H. Tabor "Using solar ponds to store power from the sun" Abstracts of Selected Solar Energy Technology.

[4] M. Collares Pereira, "Design and performance of a novel non-evacuated 1.2X CPC type concentrator" Proceedings of ISES, Intersol, Montreal, 1985.

[5] C.F. Kooi, "The steady state salt gradient solar pond" *Solar Energy*, Vol. **23**, pp. 37–45, (1979).

[6] A.E. Clark, W.C. Dickinson, "Shallow Solar Ponds" *Solar Energy Technology Handbook*, Edited by W.C. Dickinson and P.M. Cheremisinoff, 1982.

[7] TECSO TEC. Solar Ltd., Valado dos Frades, Nazare, Portugal.

[8] P. Maycock, "Photovoltaic technology progress and industrial prospects" Photovoltaic Energy Systems Inc., 2401 Chilob LA., Alexandria, VA 23308, USA.

[9] Ari Rabl, *Active Solar Collectors and their Applications* Oxford University Press, 1985.

[10] J. Hull, "Stability and economics of solar ponds using ammonium salt" Proceedings ASME Solar Energy Division Conf., Anaheim, 1986.

[11] C.E. Nielsen, "Non convective salt gradient solar ponds" *Solar Energy Technical Handbook*, Edited by W.C. Dickinson, P.N. Cheresmisinoff, 1982

[12] M. Collares Pereira, A. Joyce and L. Valle, "Salt gradient solar pond for greenhouse heating application" Proceedings ISES Intersol, Montreal, 1985.

[13] G. Lesino, Universidade de Salta, Salta, Argentina.

[14] J. Doria, M.C. de Andres, J. Yguero, A. Alonso and M. Collares Pereira, "Applicacion de los Estanques Solares a: (1) Purificacion de $NaNO_3$; (2) Obtencion de KNO_3" to be presented at the 3rd ISES Iberian Solar Energy Congress.

[15] Ari Rabl and Carl E. Nielsen. "Solar ponds for space heating", *Solar Energy* 17 (1985) 1.

2.5 Discussion Following Manuel Collares Pereira's Presentation

Faiman: Does the unique storage capability of a solar pond not place it in a special category vis à vis alternative technologies? You have not emphasized this point.

Gordon: To be more specific, perhaps, in your economic comparison of the various technologies you appear to ignore the greatly increased capacity credit for pond-generated electricity compared with the photovoltaic variety.

Collares Pereira: It is important not to get the storage advantage out of proportion when one is considering power generation. From the utility view-point 24 hours of storage is the same as no storage. Utilities require a few hours of storage in order to shift the output peak of the solar system. In my calculations I have assumed that all of the power produced by both kinds of system in a day is consumed the same day. In principle I could have included a modicum of battery storage in the photovoltaic economics but this would not radically alter the picture. In a stand-alone system this would clearly not be true, for in such cases, several days-worth of expensive storage are necessary for PV systems and here the advantages of a pond are apparent.

Weiner: It is certainly true that a solar pond power plant has extra capacity credit, but are the kWh in your table net or gross? Conventional power plants usually consume about 5% of the power produced. I would like to know what the corresponding parasitic requirements are for solar pond power plants.

Collares Pereira: For my calculations I used Ormat's published 30% figure for parasitic losses. Thus my discussion represents net power.

Doron: In theory we believe this figure to be about 25–30%. It is considerably higher at Bet Ha'aravah because not all of the planned heat exchangers were installed. In electrical figures one should expect about 20–22 kWh/m^2/year from solar ponds.

Roy: Because all of the competing technologies appear to present the same order-of-magnitude economics in your table, I would be interested to hear some more details regarding how you performed the calculations. I am particularly interested to learn how your pond experience compares to the published results of Ormat.

Collares Pereira: First, the fact that "all" competing technologies turn out to have the same order-of-magnitude economics is not the whole story. I have deliberately removed from consideration all solar technologies (shallow solar ponds, flat plates, etc.) which would have orders of magnitude worse economics.

Secondly, regarding the prices for the various kinds of technology, I reviewed the literature and adopted values typical of those one finds used by the experts. I have made no attempt to pass judgment on the degree to which these prices may be realistic or otherwise.

Thirdly, regarding the physics of each kind of system, I have correctly taken into account the proper average radiation at the surfaces of all collector types — including the relevant thresholds. Regarding the pond calculations, I was delighted to discover, upon my arrival in Israel, that the estimates I calculated for the Bet Ha'aravah pond while in Portugal are extremely similar to the figures presented here at last year's workshop by Mr. Doron — in Hebrew and consequently unknown to me at the time I did my calculations! I thank Prof. Zarmi for subsequently translating those results into English for me!

As for my own pond experience, the results are not really comparable. Mine is a low temperature pond (for thermal energy production) and located in an agricultural area. These factors enhance algae growth and result in a transparency of only 0.26 compared to Ormat's 0.38.

Zarmi: How do you define utilizability for a pond?

Collares Pereira: I compute the difference between the incoming radiation and the sum of heat losses and energy drawn. For purposes of simplification, I assumed a constant temperature difference all year. This would correspond to something like 85°C–35°C in summer, and 70°C–20°C in winter.

Tabor: Notice that for ponds one can ignore threshold factors. For, unlike a flat-plate collector, a pond loses heat all the time — day and night. So both solar input and heat losses can be replaced by integrated constant values.

Editor's comment: Dr. Harry Z. Tabor (1917–2015) was the scientist who first conceived, *inter alia*, the concept of a solar pond power plant (*vide* Dr. Collares Pereira's reference (3) and all the early literature on the subject) and who, together with the Ormat company, brought it to fruition. He was a regular attendee at the Sede Boqer Symposia on Solar Electricity Production, and one of his after-dinner talks, pertinent to this particular subject, is reproduced in Chapter 4 of this book.

Chapter 3

1988

3.1 Editor's Foreword

For the third Sede Boqer Symposium we were fortunate to attract 3 keynote speakers. Jan Kreider was the first independent engineering specialist to have gained access to one of the large solar-thermal power plants that Luz Corporation constructed in California. During the several days he spent at SEGS V he was able to study the performance of this system in real time. His presentation at Sede Boqer was accordingly a unique state-of-the-art document, unadorned with the advertising "hype" that conventionally accompanies presentations from personnel associated with most manufacturing companies.

Roland Winston was the first scientist to draw the attention of solar energy researchers to the important fact that optical devices which produce a high-quality image of the Sun do so at the price of missing large amounts of the energy that arrives from that source. What subsequently became known as the "Winston concentrator" was the first of a new generation of practical devices that have since been fashioned in order to obtain optical concentration close to the thermodynamic limit. His Sede Boqer review explains in quantitative manner how this can be done.

Fred Treble (1916–2010) was one of the most highly qualified and respected photovoltaics specialists associated with the early efforts of the European Union to investigate the performance of PV systems throughout the continent of Europe. His Sede Boqer review, which will be included in Volume 2 of this series, gives an indication of the wide range of terrains and climates in which the performance of those early systems was studied.

3.2   Performance of a Large Solar-Thermal Electric Power Plant in the USA (Prof. Jan Kreider)

A keynote lecture presented by Professor Jan Kreider (Joint Center for Energy Management, University of Colorado, Boulder, CO 80309, USA).

3.2.1   *Introduction*

The purpose of this lecture is to present independent observations on the performance of the largest and newest solar thermal plant for power production in the world. To the author's knowledge this is the fifth of five Solar Electric Generating Systems (SEGS's) located in the high desert of Southern California. This 30 MW plant is one of five solar thermal plants using steam-driven electrical generators to produce and sell power to the Southern California Edison (SCE) grid under five separate 30-year contracts. SCE is a large utility in the American Southwest selling power to the metropolitan Los Angeles, California area and surrounding cities.

This paper will describe the most recent SEGS plant and report on measured performance during 1987. Comparisons of power delivered to predictions, and other observations, will be made. Lessons learned will be summarized. A day-long site visit was made by the author less than a month ago to acquire the most recent possible information. The majority of data used in this report were generously and openly provided by Luz Engineering Corporation (LEC), the operator of the SEGS facilities. Luz International, Ltd. is a California corporation with corporate divisions in Israel (Luz International Israel — plant engineering, collector production and R&D) and the US (LEC — project engineering, plant operation and maintenance; Luz Development and Finance Corp — solar power development and finance).

A number of comparisons and overviews of thermal and other plants for producing power from the sun have been presented [1,2,3]. In addition, a number of reports and papers have been issued

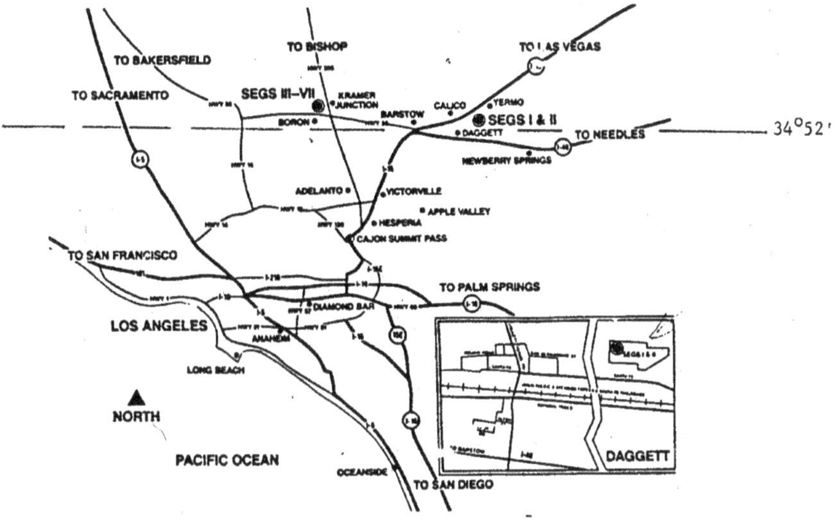

Figure 3.2.1:   Location of SEGS power plants

on specific projects in the past [4, 5]. Standard texts describe the important parameters needed to calculate the performance of concentrating collectors and thermal power plants [6,7].

3.2.2   *Overview*

In the early 1980's Luz Engineering Corporation (LEC) made application to the Federal Energy Regulatory Commission (FERC) to sell power under the US PURPA (Public Utilities Regulatory Policy Act) to SCE. The first of the SEGS plants (14 MW) began operation in 1984. Four more plants (SEGS II–V) have been added to the SEGS system as of this time. Each of the plants, after the first, was limited to 30 MW due to restrictions placed by the then current version of the Public Utilities Holding Company Act.

The locations of the SEGS plants are shown in Fig. 3.2.1. The three most recent plants (SEGS III–V) are near Kramer Junction, CA, 110 km northeast of Los Angeles. The elevation of this site is approximately 600 m; the latitude is approximately 35 degrees north.

Historical weather data for nearby Daggett, CA are presented in Table 3.2.1. This desert climate has a mean annual clearness index of

Table 3.2.1:   Insolation and temperature data for Daggett, CA: Latitude 34°52'N; Longitude 116°47'W; Elevation 588 meters

	Average monthly temperature [°C]	Global $\langle K_T \rangle$ clearness index	Daily direct normal irradiance [kWh/m^2]
Jan	8.50	0.586	4.87
Feb	11.11	0.617	5.51
Mar	13.72	0.668	6.60
Apr	17.94	0.713	8.01
May	22.38	0.731	8.69
Jun	26.72	0.751	9.39
Jul	30.72	0.721	8.76
Aug	29.72	0.716	8.32
Sep	26.22	0.701	7.59
Oct	20.05	0.661	6.65
Nov	13.05	0.612	5.52
Dec	8.88	0.580	4.83
Ann	19.11	0.687	7.07

Table 3.2.2:   Principal characteristics of the SEGS plants I–VII

Plant	Year on-line	Nominal capacity [MW]	Gross aperture area [m$^2 \times 10^3$] (collector type)	$T_{C,O}$ [°C]	Efficiency [%] Solar	Gas
I	1985	14	83 (LS-1,2)	307	31.5	—
II	1986	30	165 (LS-1,2)	315	29.4	37.3
III	1987	30	204 (LS-2)	349	30.6	37.4
IV	1987	30	204 (LS-2)	349	30.6	37.4
V	1988	30	233 (LS-2)	349	30.6	37.4
(VI)	(1989)	30	188 (LS-2)	390	37.5	39.5
(VII)	(1989)	30	183 (LS-3)	390	37.5	39.5

0.69, and mean ambient temperature of 19°C. Average daily, direct normal radiation varies between 9.4 and 4.8 kWh/m^2 [8].

References [9, 11] summarize the earlier plants and their operating experience. Table 3.2.2 lists key characteristics of the five present plants and the next two plants planned for construction this year. The three rightmost columns indicate the improvement in turbine efficiency and collector outlet temperature which have occurred through 1988 and are planned for the next two plants. More will be said about future plants later.

According to the utility contracts for SEGS III–V, revenue payments for power produced depend on the rate of production, the time of production and the aggregate energy production. Energy payments are based on MWh production irrespective of time of day. Capacity payments are made based on rate of energy production and time of day (the peak, mid-peak and off-peak periods are described below). Full (100%) capacity payments are achieved if production is at 80% of rated capacity level during the peak period. If actual capacity exceeds 85% of the rated capacity, bonus capacity payments are made.

If solar heat is unavailable, a gas boiler is used to produce steam. The amount of gas which may be used for this purpose is limited by FERC to 25% of the total input energy on an annual basis. In addition, Luz warrants to the third party, private owners of the plants, that certain levels of revenue from power production will be met. This warranty is adjusted yearly based on insolation. As a result, the SEGS plants must operate at a high level of reliability for an extended period. A new level of collector design, manufacture and operation is required. Likewise, the power generator must have a high reliability level. The balance of this paper will describe the details of the most recent plant, SEGS V, put on line early in 1988. Table 3.2.3 summarizes the key characteristics of SEGS V. The contents of this table will be described shortly. Since annual performance data are not yet available for this plant, data from SEGS IV will be the basis for performance comparisons to follow.

3.2.3 *Collector design*

The latest generation of collectors used at SEGS V is known by the acronym LS-2 (i.e., Luz System 2). This device is a north-south, horizontal axis parabolic trough collector with a concentration ratio of 71×. The 5m by 47m aperture captures direct beam radiation which is in turn reflected from back-silvered glass reflectors to the collector absorber tube; the mirror rim angle is 80 degrees. This

Table 3.2.3:   Nominal plant characteristics of SEGS V

LS-2 Collectors		992 units
Total Aperture		$233 \times 10^3 \, \text{m}^2$
Design Outlet Temperature		$349°\text{C}$
(Controlled by Flow)		
Annually Averaged Efficiency		50%
Optical Efficiency		74%
Heat Transfer Fluid	Monsanto	VP-1 (Freezes at $13°\text{C}$)
Ground Cover Ratio		2.5
Design Gross Output		33 MW
Design Net Output		30 MW
Dual Inlet Turbine	Solar Port	$327°\text{C}$, 4.34 MPa
	Gas Port	$510°\text{C}$, 10.5 MPa
Heat Rate	Solar	11,766 kJ/kWh
	Gas	9,642 kJ/kWh
Efficiency	Solar	30.6%
	Gas	37.4%
Annual Output	Net	92,400 MWh

receiver is constructed of stainless steel whose outermost surface is plated with black chrome with a nominal absorptance of 94% and emittance at $300°\text{C}$ of 24%. The absorber is surrounded by an evacuated (0.0001 torr) high transmittance glass tube. The optical intercept factor [6] of this collector is 97% [9]. Nominal collector optical efficiency is 74% with peak conversion efficiency of solar flux to heat of 66%. The important characteristics of the LS-2 collector are shown in Table 3.2.4. The LS-3 collector whose characteristics are also given in the table will be described later.

Tracking of the sun is accomplished by a single motor which positions the collector assembly to within an accuracy of 0.1 degrees. Approximate alignment is accomplished by feedback control from the central plant operating system. However, final precise control is accomplished by the local tracker at each collector. The local sun sensor is of novel design using a convex lens which creates an image on two solar cells at the sensor's focal plane. A 15,000:1 gear reduction provides sufficiently small position steps, to assure required tracking accuracy. Collectors near the edge of each field where wind effects can be significant have additional physical features added to the collector to dampen wind buffeting. The field is stowed at winds above 16 m/s.

Table 3.2.4: Comparison of the Luz LS-2 and LS-3 collectors

	LS-2 (SEGS III-VI)	LS-3 (SEGS VIII-...)
Aperture [m^2]	235	545
Concentration Ratio	71	82
Rim Angle	80°	80°
Receiver Material	Stainless Steel with Black Chrome	CERMET
Optical Efficiency	0.74	0.77
Receiver Absorptance	0.94	0.97
Receiver Emittance	0.24 (300°C)	0.15 (350°C)
Tube Transmittance	0.95	0.97
Mirror Reflectance	0.94	0.94
Peak Efficiency	66%	68%
Structure	Torque Tube	Central Truss
Tracker Feedback	Inclinometer	Encoder

During periods of solar outage the collector field is also stowed — face down.

3.2.4 *Collector field and turbine*

The 233,000 m^2 of solar collectors at SEGS V are arranged in 70 loops of 14 collectors each. Each of the 70 loops is connected in parallel to the main field header with insulated piping. Each tracking collector is connected to the fixed field piping by a flexible stainless steel pipe approximately one meter long. The ground to cover ratio at this plant is 2.5 (i.e. ground area to aperture area ratio).

Collectors are cooled by an organic working fluid manufactured by Monsanto Industrial Chemicals, Inc. The fluid is known as Therminol VP-1^{TM} and is a blend of diphenyl oxide and biphenyl materials. This fluid was selected owing to its stability at elevated temperatures. However, the cost for this capability is US$2.50 per liter. It has a relatively low vapor pressure (relative to water) of 540 kPa at the design point of 349°C but has a specific heat only a little more than half that of water. The density is about 75% that of water but the freezing point is 13°C. This latter property requires circulating and heating the fluid during the coldest parts of the winter. Owing to the expense of this fluid, thermal storage is not feasible. The nominal flow

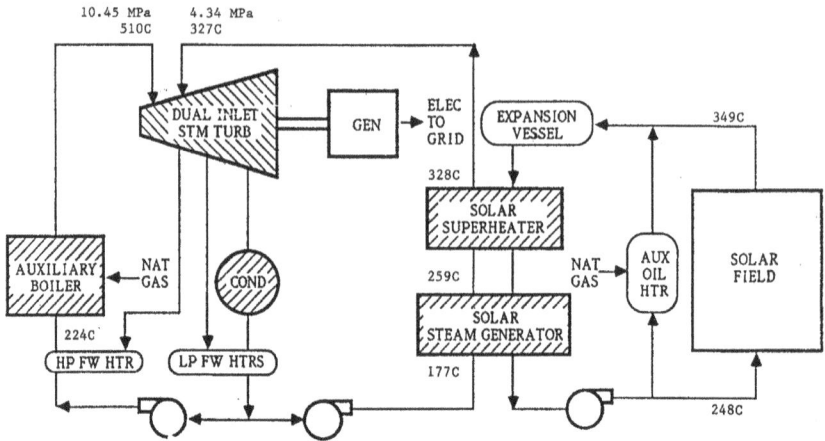

Figure 3.2.2:   Schematic diagram of the SEGS III-V plants

rate for a 30 MW plant such as the SEGS III-V family is 550 liter/s
(0.00233 l/s/m²).

Figure 3.2.2 is a schematic diagram of the SEGS V plant showing
the arrangement of collector field, steam generators and the dual-port
turbine-generator. Hot fluid produced in the solar field is first passed
in counter-flow through the steam super-heater and then through
the boiler. Superheated steam at 327°C and 4.3 MPa enters the dual-
port turbine. Exhaust steam is used for feed water heating and then
condensed in a standard cooling tower. The nominal efficiency of the
Westinghouse turbine at this condition is 30.6%.

In the event that insufficient solar heat is available, steam is
produced in an auxiliary, conventional gas boiler at 510°C and
10.5 MPa. This higher pressure/temperature steam is introduced into
the turbine at a separate port and expanded to the same condition as
for the solar produced steam. Turbine efficiency under the gas fired
condition is 37.4%. Gross power output of the turbine is 33 MW
whereas the net output is rated at 30 MW.

3.2.5   *Performance and operational results*

Revenues paid by SCE are highest for power produced from solar heat
during the summer peaking period shown in Fig. 3.2.3. This period

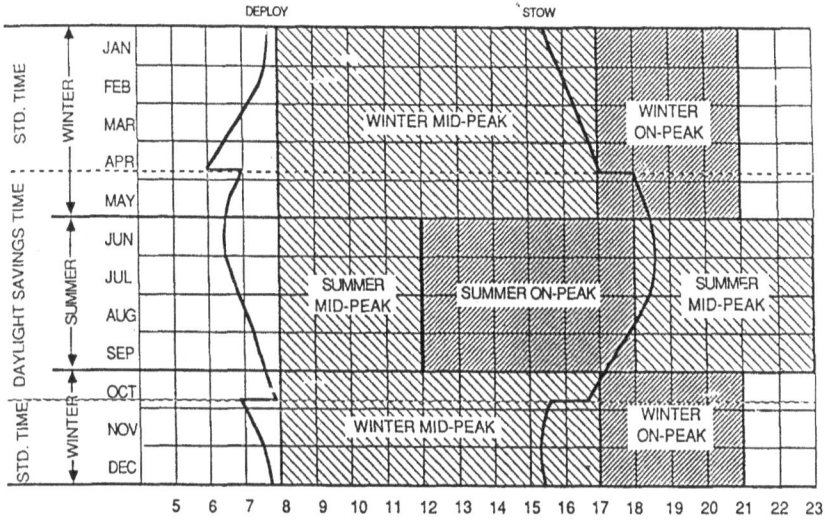

Figure 3.2.3:   Contract rate periods for the SEGS III-VII plants

extends from noon until 6:00 pm between June 1 and September 30. Sixty percent of the plant's revenue was produced from peaking period production in 1987. The figure shows that little solar-based power can be produced during the winter peaking periods; gas is used extensively during the winter peak. The winter peaking period has been deleted from the Luz/SCE agreement for three years as of January 1, 1988. The definition of peaking periods and rates can be changed in the future with agreement of the California Public Utilities Commission. At off-peak periods and during weekends, little gas is used; the turbines are operated primarily with solar produced steam.

Table 3.2.5 is a summary of performance projections based on weather data for the area assembled by WEST Associates, a consortium of California Utilities. (These data are about 5% greater than TMY data for nearby Daggett, CA, used elsewhere in this article.) The data for SEGS I are included for comparative purposes. The small difference between SEGS IV and V performance has to do with slight solar field differences between the two. Data presented hereafter are measured, not simulated.

Table 3.2.5:   Comparative annual plant performance trends

Plant	$T_{C,O}$ [°C]	Solar [GJ/m²]	Electricity [MWh/m²]	kWh/GJ	Solar-to-Electricity Efficiency Peak	Average
I	307	3.6	0.34	95.8	22%	10%
IV	349	3.7	0.40	108	19%	11%
V	349	3.6	0.40	111	19%	11%
(VII)	393	4.1	0.54	132	24%	14%

3.2.5.1   *SEGS IV performance*

Figure 3.2.4(a) shows the fraction of solar collectors in operation for the 12 months of 1987. Even though 1987 was the startup year for this plant (with reduced contractual performance requirements), the collector availability met the mature year criterion of 97% availability for about half of the year. A "mature" year is any year after the startup year.

Figure 3.2.4(b) shows that energy production met or exceeded the amount warranted to investors during a mature year for a significant part of this startup year. September and October experienced lower than average insolation and startup-year limitations on natural gas availability. Boiler repairs in May reduced plant output temporarily. The key conclusion from this figure is that performance is better than projected for a startup year and about as predicted for a mature year.

Cumulative energy production is presented for peak period (Fig. 3.2.5(a)) and total production (Fig. 3.2.5(b)) bases. The lower curve in each graph is the startup year projection adjusted for 1987 insolation. The upper solid curve is the mature year projection. The points on the third curve represent actual, cumulative performance of SEGS IV. As before, performance is well above startup year projections — 150% of startup requirements.

Figure 3.2.6 shows the fraction of gross electrical production attributable to solar heat. As designed, little auxiliary gas heat is used during the peak summer season. The only contractual restriction on gas use is the FERC 25% rule which has been

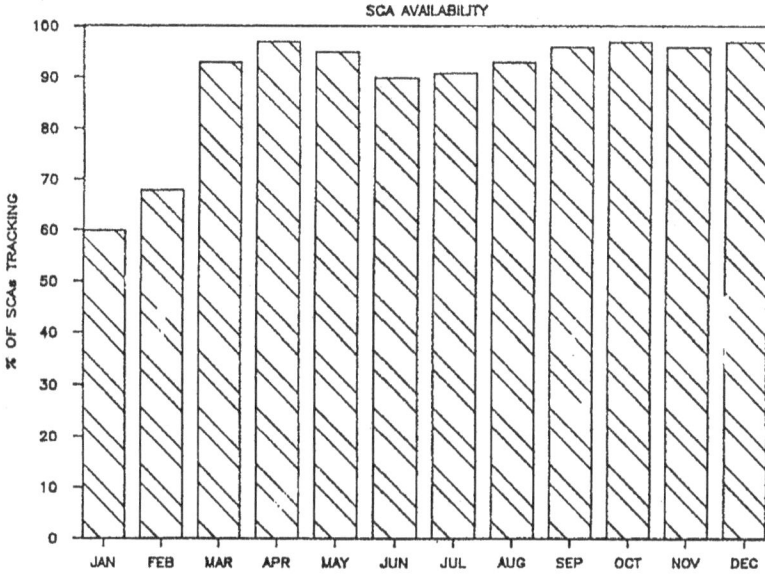

(a)

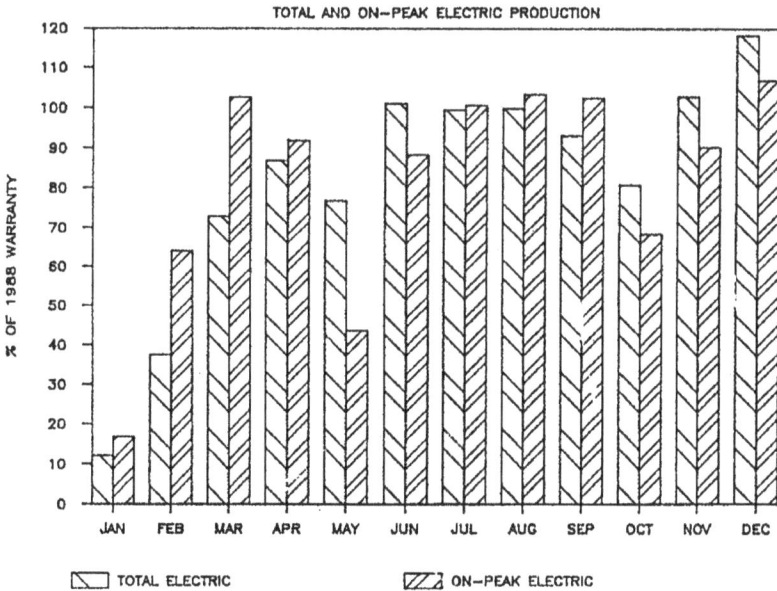

(b)

Figure 3.2.4: (a) SEGS IV collector availability during the year 1987.
(b) Measured total and on-peak performance of SEGS IV during 1987

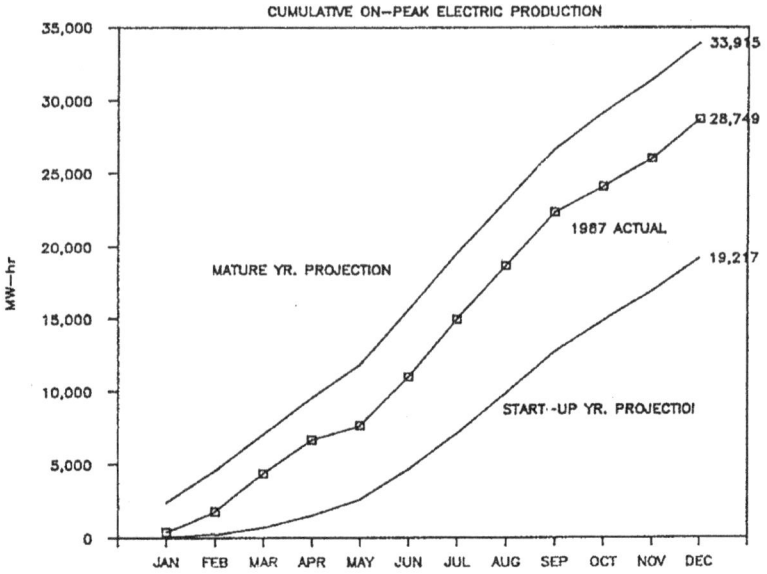

(a)

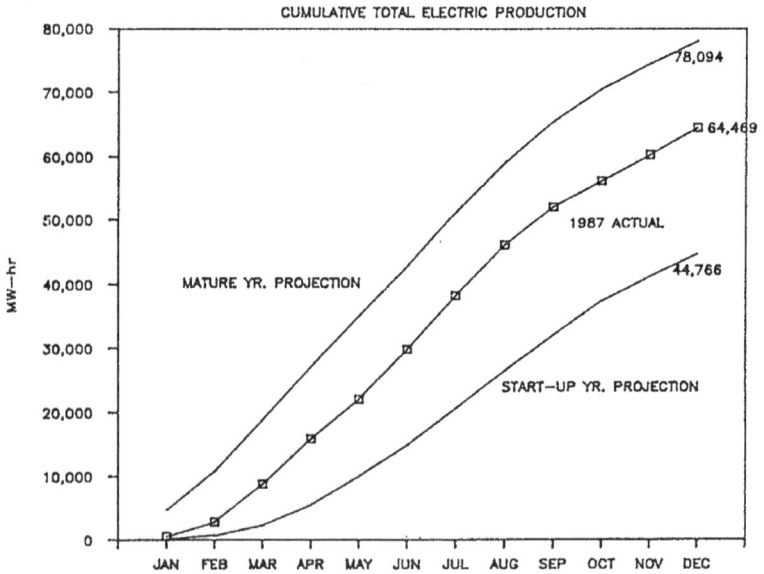

(b)

Figure 3.2.5: (a) Cumulative on-peak production of SEGS IV during 1987. (b) Cumulative total production of SEGS IV during 1987

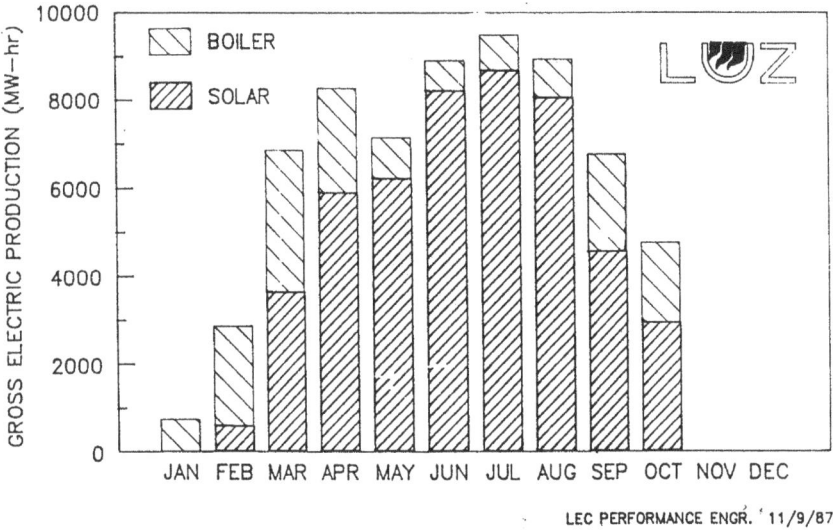

Figure 3.2.6: SEGS IV solar and boiler gross electric production during 1987

Table 3.2.6: SEGS IV performance data for startup year 1987

Solar Flux Availability		90% of Nominal
Solar Field Availability		92.5%
Solar Electric Output	Gross:	52,200 MWh
Plant Output	Net:	28,750 MWh
	Net Total: 64,470 MWh	
Gas Consumption		303,280 GJ
Contractual Performance	Fraction of Startup Year: 150%	
	Fraction of Nominal Year: 85%	

interpreted for the SEGS facilities to mean that

annual gas use/(annual gas use + solar input/0.8) < 0.25

Figure 3.2.6 data have been totaled for all of 1987 in Table 3.2.6 to present an annual summary of SEGS IV performance. Insolation was 90% of the design year during 1987. Gross solar production amounted to 52.2 GWh with net plant output of 64.5 GWh including gas firing periods. Gas consumption was within the 25% FERC limit.

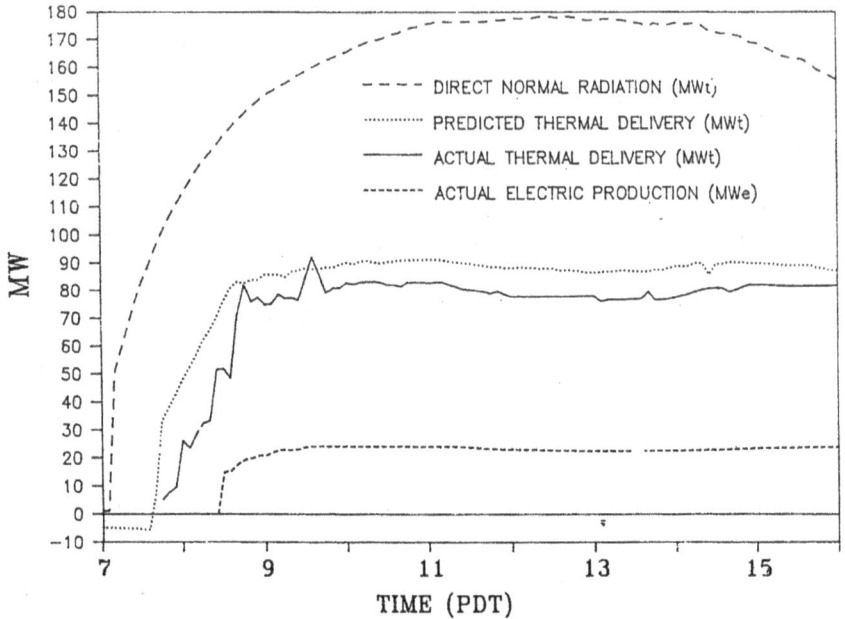

Figure 3.2.7:   Hourly performance of SEGS III for 20th September 1987

Data measured during a sunny fall day are plotted in Fig. 3.2.7 for SEGS III. Sunrise occurred shortly after 7:00 am daylight time with useful power production starting Just before 8:30. Power was produced at the 25 MW level continuously during the day. The dotted line is a performance prediction based on a Luz thermal model [12] driven by insolation data for this day in September.

3.2.5.2   *Comparisons with rules of thumb and simplified methods*

There are a number of rules of thumb and simplified design approaches used by solar designers to estimate component and system performance during the early stages of design. A few of these will be briefly noted in this section.

One of the conclusions of finite time thermodynamics [13] is that the efficiency of a real heat engine including finite rates of heat transfer from and to heat reservoirs is given by

$$\text{Efficiency} = 1 - \sqrt{(T_{cold}/T_{hot})}$$

Use of this simple result to check the SEGS IV turbine efficiency, when solar fired, gives a value of 30.2%. The rated value of the turbine at design conditions with solar firing is 30.6% according to the manufacturer. Checks for other conditions and other SEGS systems give similar results. Although not a solar-related conclusion, this simple check indicates that the turbine efficiency values quoted are realistic.

Another check which can be performed is a prediction of the nominal annual collected energy by the solar field. The simplest of the short methods is the utilizability approach [14]. Using climatic data from Table 3.2.1 and collector specifications from Table 3.2.4 and an estimated collector heat loss conductance of $0.34\,\text{W/m}^2/\text{K}$, one estimates approximately $4.2\,\text{GJ/m}^2$ of annual collected heat for the SEGS IV field. This provides another independent check on the design calculations.

3.2.5.3 *Collector azimuth study*

Since the peak revenue producing period for the SEGS IV plant (and all other plants) is during summer afternoons, it may be advantageous to position the axes of future SEGS plants to favor the afternoon sun. A study made by the author of incident solar flux during the summer peak indicated that about two percent more flux is intercepted by a collector field oriented between 15 and 20 degrees toward the west instead of due south. (Fig. 3.2.8). This study based on hourly TMY data from nearby Daggett, CA is of course only preliminary. Revenue loss at off peak periods should be compared to revenue enhancement (or equivalently reduced area needed to produce the same revenue) during peak periods.

3.2.6 **Financial considerations**

The cost of the three most recent plants averaged US$110M. On a gross output basis this amounts to $3,400/kW. If the nominal energy

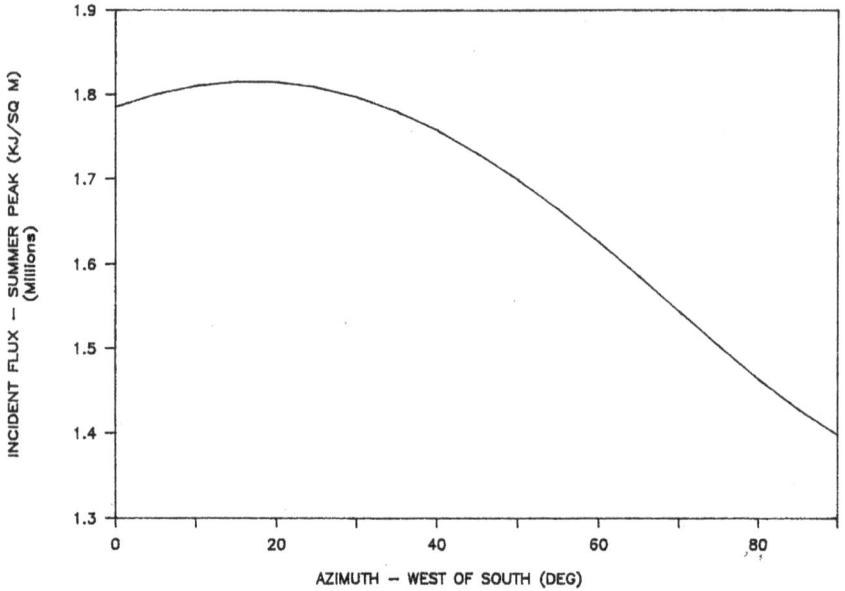

Figure 3.2.8:   Potential gain from a westward shift of trough axis

rate of 92,400 MWh/y is achieved on the average, the cost of electric power will be approximately \$0.133/kWh. With improvements and larger unit size planned for future plants, this price is expected to drop to \$0.06/kWh (1991, 80-MW plant).

Figure 3.2.9 is a plot of plant cost per kW vs time. The leftmost two points are based on actual costs. The rightmost points are projections.

Figure 3.2.10 is a similar plot but for the collector field only. With the advent of the next collector generation (see below) costs continue to drop slowly but performance is better. Since the first collectors were built five years ago, collector cost has dropped by a factor of eight.

The SEGS plants have been financed under various tax scenarios [11] the net result of all of which was an internal rate of return to the private, third-party investors of 14–15% per year. Future plants are planned to have essentially the same rate of return.

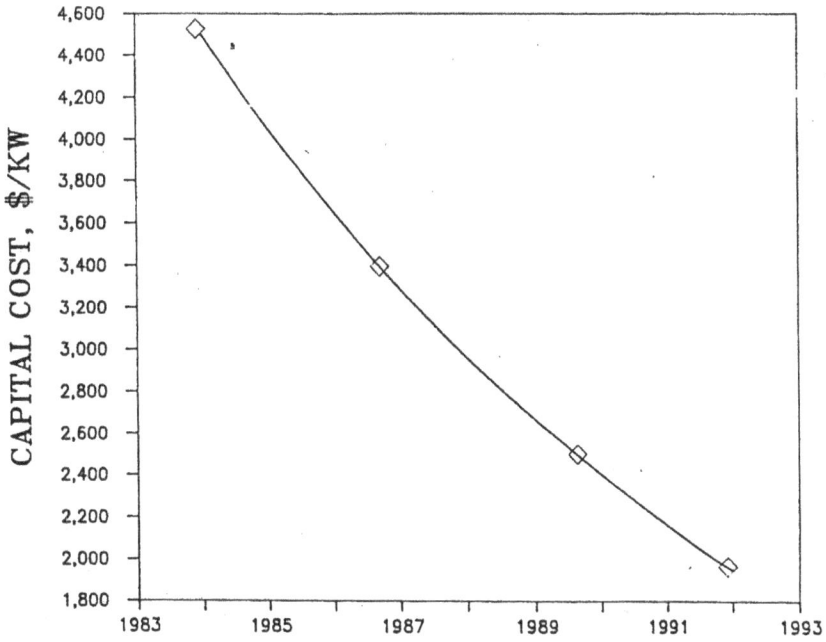

Figure 3.2.9:   Capital cost trends of SEGS plants

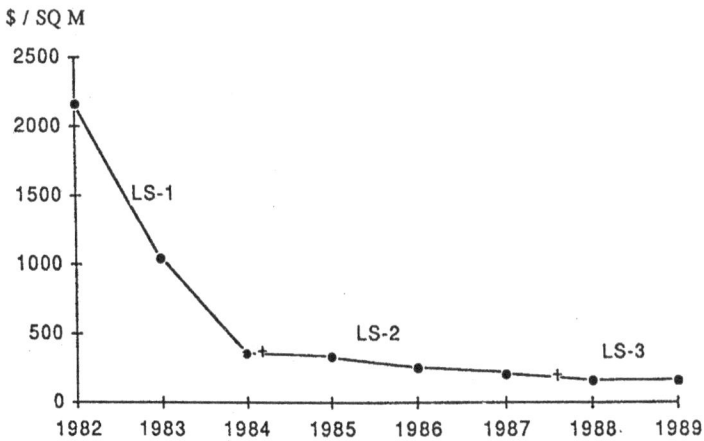

Figure 3.2.10:   Luz solar field cost trends

From the viewpoint of the SCE utility, the SEGS plants displace their full rated capacity of 30 MW per plant of conventional generating capacity.

3.2.7   *Lessons learned and future improvements*

Most of the important lessons learned have been documented elsewhere [9,10,11]. Problem areas in the field were primarily in the flexible connections between collectors at early plants and the rigid pipe network. These have been solved by use of a new vendor for the flexible hose fabricated from stainless steel. Collector position sensors (inclinometers) have caused problems and are being replaced by tracking axis shaft encoders. Heat transfer pumps and cooling tower control have also been improved. Fluid leaks were a problem in earlier systems but have been reduced significantly.

Other components have been very reliable. Breakage of glass tubes and mirrors has been well below one percent. The most recent plants have been started without significant problem due to experience with prior plants and thorough training programs.

Future improvements are centered on a new collector design. These will be detailed in another lecture at this Symposium. Table 3.2.4 shows the key collector differences. An improved absorber will be used with reduced emittance, higher absorptance and higher concentration ratio. The new LS-3 model is twice as long as the current LS-2 collector thereby reducing the number of local controllers, flexible hoses and temperature sensors. An improved structural system is used for better dimensional control and reduced cost. As a result of these improvements; efficiency will increase by three points. Because of improved performance, fluid outlet temperature will be higher and turbine efficiency greater as summarized in Table 3.2.2 and Fig. 3.2.11.

3.2.8   *Summary*

A large solar thermal electricity plant has been constructed in the Southern California Desert. During the first year of operation the newest, 30-MW module performed 50% better than predicted.

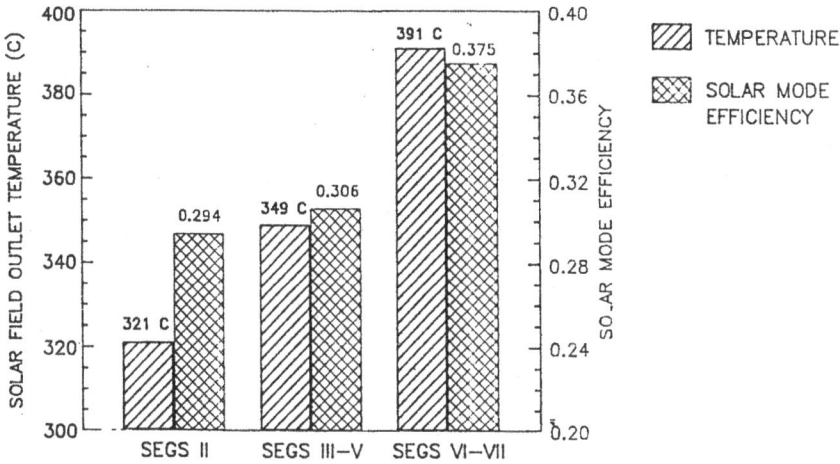

Figure 3.2.11: Effect of solar field outlet temperature on power cycle efficiency

Significant improvements have been made in reliability, efficiency and cost reduction during the past five years. The combination of these characteristics combines to produce electricity reliably at a cost which can be sold at a price returning 14–15% annually to the plant's owners.

3.2.9 *Acknowledgments*

I wish to thank Dr. David Kearney and Mr. Henry Price of Luz for openly and generously showing me the SEGS plants and the performance data for the plants. Dan Jaffe of Luz described his performance model for the SEGS plants.

3.2.10 *References*

[1] D. Faiman, Ed., "Proceedings of the First Sede Boqer Symposium on Solar Electricity Production", 23–24 February, 1986.

[2] D. Faiman, Ed., "Proceedings of the Second Sede Boqer Symposium on Solar Electricity Production", 25–26 February, 1987.

[3] F. Kreith, R. Davenport and J. Feustel, "Status review and prospects for solar industrial process heat", *Journal of Solar Energy Engineering* 105 (1983) 385.

[4] D.L. Larson, "Performance of the Coolidge solar thermal power plant", *Journal of Solar Energy Engineering* 109 (1987) 2.

[5] L.R. Bush, Solar total energy system progress reports, aerospace corp., ATR-77 (7692-01)-1 and -2, 1977.

[6] J. Kreider, *Medium and High Temperature Solar Processes* Academic: Press, New York, 1978.

[7] W. Stine and R. Harrigan, *Solar Energy Fundamentals and Design* John Wiley, New York, 1985.

[8] C. Knapp and T. Stoffel, "Direct Normal Solar Radiation Data Manual", SERI Report No. SERI/SP-281-1658, Golden CO 1982.

[9] D. Kearney and H. Price, "Overview of the SEGS Plants", Proceedings of the 1987 ASES Annual Meeting, p.69, Portland, OR 1987.

[10] D. Jaffe, S. Friedlander and D. Kearney, "The Luz solar electric generating systems in California", *Proceedings of the ISES Congress*, Hamburg, Sept., 1987.

[11] D. Kearney, H. Price, Y. Gilon and D. Jaffe, "Performance of the SEGS plants", to be presented at the ASME Solar Energy Division Annual Solar Conference, Denver, CO, April, 1988.

[12] D. Jaffe, private communication, March, 1988.

[13] M. J. Ondrechen, *et al*, "Maximum work from a finite reservoir by sequential Carnot cycles", *Amer. J. Phys.* 49 (1981) 681.

[14] J. Sueker and J. Kreider, "Solar utilizability equations — An update", Proceedings of the 1987 ASES Annual Meeting, p. 387, Portland, OR, 1987.

3.3 Discussion Following Jan Kreider's Presentation

Question: You have stressed the closeness of SEGS V's performance to that expected from a mature technology. How well does SEGS II perform?

Kreider: The initial performance predictions for both SEGS I and SEGS II turned out to be somewhat optimistic. Those fields were then upgraded, via the addition of more collectors, so as to meet their contractual requirements. Thus all SEGS systems perform as examples of a mature technology.

Roy: SEGS I originally contained 70,000 m^2 of collectors. I had understood that Luz had subsequently discovered the need to add a further 20,000 m^2, making 90,000 m^2 in all. Yet your table indicated only 83,000 m^2 for this system. What is the source of this discrepancy?

Kreider: The original field contained model LS-1 collectors. The LS-2 second generation collector is more efficient and only about 13,000 sq.m of the latter turned out to suffice in order to bring the system performance up to contractual requirements.

Question: How does the 10% figure you quoted for the overall solar-to-electricity efficiency of these solar-thermal plants compare with the performance of photovoltaic power stations?

Roy: I have studied that question. The Hesperia photovoltaic system in California produces AC power at about 6% efficiency. It seems that at the present stage of development, plants of the solar-thermal variety are slightly more efficient and slightly lower in cost than their photovoltaic competition.

Faiman: But you have to be careful here. The 10% Luz figure is relative to the beam component, whereas your 6% Hesperia figure is relative to the total insolation incident on the collector plane. These two efficiencies would not seem so different if both were referred to the total incident radiation.

Kreider: In some sense the whole concept of efficiency is somewhat of an artifact. It represents some kind of intermediate number which tells you how well a specific system is converting solar energy to electricity. It is not a very meaningful parameter for assessing the relative performance of two different systems.

Treble: Could you say something about the operation and mainte-nance costs of these plants?

Kreider: Maintenance costs amount to about 2% per year of the system cost. A staff of approximately 100 personnel attends to the three plants, performing cleaning, monitoring, repairs, etc.

Glueckstern: How do you explain the fact that these line-focus systems, which are commonly regarded as being inferior to the higher temperature technologies, are so commercially successful?

Kreider: I have visited and inspected many solar sites during the past decade but SEGS III-V is by far the most professionally operated system I have ever seen. For example, these plants are run by power plant people, and they have been designed so as to ensure that maintenance is a simple routine.

As for the higher temperature technologies, there does not seem to have been sufficient commercial push to improve them. The central receiver system at Barstow, for example, was built at enormous cost but was never subsequently developed or improved.

One higher temperature technology that did receive some commercial interest was the point-focus system that LaJet built in 1985, about 40 miles from the SEGS V site. Although I have not inspected the LaJet system I am told that there have been serious problems: The mirrors have delaminated, and the liquid metal absorber has proved to be ineffective, resulting in the need to redesign the power units. I have heard that the system is not currently operational and that LaJet are now looking for additional investors.

By comparison, Luz invested heavily in the design of a solar collector field that would be both well engineered and that could feed conventional power generating equipment. This strategy seems to have paid off for the time being.

Glueckstern: How do you view the future of line-focus systems relative to the potential advantages of the higher temperature technologies?

Kreider: Regarding central receivers, I do not see any progress for the time being, and point-focus systems seem beset with materials problems. What the future will bring is impossible to know. But for the time being line-focus systems are a tried and tested concept.

The fields are modular in that you can add collectors or shut others down for repair, and the SEGS plants have better load factors than was ever the case for the Barstow central receiver.

Weiner: As a power plant man I am greatly surprised at the availability figures for first year operation. How can it be that a relatively new kind of technology, spread out over a large area, can display superior first year performance figures to a conventional fossil fuel plant?

Kreider: I am not a power plant expert so I am not familiar with what is involved with the start-up of a conventional plant. But I should guess that there are greater economic incentives to get solar plants up to full performance as soon as possible.

Roy: Following on from Dr. Weiner's question, I'd like to enquire what the capacity factor of these SEGS plants is, i.e. How many hours per year do they give at full load equivalent?

Kreider: Because of their gas backup they operate like any conventional 30 MW rated plant. The utility requires 30 MW during their peak load periods and they get it. They are not concerned with how the 30 MW are generated. Perhaps I should elaborate slightly on the economic incentives involved.

The plant only needs to operate at 80% of its rated capacity (e.g. 24 MW for a 30 MW plant) in order for the owner to receive full payment. Above this figure — which might typically be anywhere in the range of 80% to 100% of the nominal power rating — the plant owner actually receives a bonus. So there is a clear incentive to keep these plants operating well.

Roy: Could you clarify the 25% limit figure on the use of backup which you referred to in your presentation?

Kreider: The actual formula is that:

$$\text{Gas backup}/[\text{gas backup} + (\text{solar}/0.8)]$$

must not exceed 0.25, where "solar" refers to the incoming beam radiation — not its thermal derivative.

Thus, for example, it would not be cost-effective for the plant owner to burn all of his gas allocation during summer peak hours because the gas limit would not enable him to produce enough revenue. What he needs to do is to ensure that the solar fraction of his plant is as high as possible.

Gordon: If there were not a 25% limit on the use of gas backup and the tax structure were to remain unchanged, would it be more worthwhile for Luz to operate on 100% gas rather than to build a solar field?

Kreider: First, there is not enough gas available for them to do that. Secondly, if there were enough, they would be a conventional utility and, as such, would not qualify for the special tax credits associated with the use of renewable energies.

Tabor: I should like to know what the local price of gas is, and what rates the utility pays for on-peak and off-peak kilowatt–hours.

Kreider: Gas costs about US$0.6/Therm. SEGS V produces electricity at a cost of about US$0.13/kWh but I do not know how much the utility pays for this power as I have not seen the contract.

Tabor: If these plants spend US$0.13/kWh in order to produce power then they cannot meet Dr. Glueckstern's criterion for competing with the marginal costs associated with fossil plants.

Kreider: This is not necessarily the case. My figure of US$0.13/kWh is simply the capital outlay divided by the energy produced. It does not include the effect of a quite complex taxation structure. [These include a 10% tax credit from the State of California, an 8% Federal tax credit and a 15-year depreciation write-off.] In point of these facts, these plants have yielded a 15% return to their investors.

Tabor: Then presumably the tax payer is providing these profits. If your collectors cost US$200/m^2 and are, say, 50% efficient, then the heat output costs about $0.20/kWh — which is very costly for thermal energy.

Roy: Most countries subsidize energy in its various forms — for example, nuclear energy. I therefore see no reason to criticize solar plants on the ground that they depend on tax subsidies. This is particularly true at the onset of a relatively new technology. It would seem to be sensible to support solar at its development stage in order to ensure that the technology will be available at a time when it may really be needed.

Gilon: Perhaps I might add a few words to help clarify the economic aspects of the Luz plants.

First, the US$0.07/kWh we pay for electricity here in Israel is relatively inexpensive by world standards. For example, in Southern California the public pays more than US$0.20/kWh for summer on-peak electricity. Thus Luz could operate at a profit even without the tax credits. Indeed, the tax credits are due to disappear in 1989, but I believe we shall continue to survive.

Secondly, the quoted figure of US$200/m^2 for the solar field includes the cost of the power block. The solar part alone is only US$150/m^2. If we could reach US$100/m^2 for the solar part we should be completely competitive with a regular utility.

Thirdly, as regards the time-of-day schedule you showed for electricity tariffs: These actually changed on January 1st 1988. There is no longer any winter evening peak credit. This actually puts the solar field geometry in even better harmony with the summer peak requirements. In fact, about 2/3 of the annual revenue now comes from summer on-peak production. Some 40% come from the four 6 h/day summer months, and about 50% of our revenue comes from just 5 days per week of operation during these 4 months. This would be considered respectable performance by most small plant operators.

3.4 Novel Optical Methods for Increasing the Power Output of Solar Conversion Systems (Prof. Roland Winston)

A keynote lecture presented by Professor Roland Winston (University of Chicago, Chicago, IL 60637, USA).

3.4.1 *Introduction*

It is a significant pleasure for me to address this conference on this occasion of my first visit to your country. I am looking forward to discussing the application of the methods and techniques of non-imaging optics to solar energy conversion technology.

The task of non-imaging optics in solar energy conversion technology is to concentrate the solar flux at the least possible "cost" in tracking tolerance, precision of components and so on. It may be appropriate to remind ourselves why we bother to concentrate the solar flux at all. As every worker in solar energy conversion quickly discovers, the ambient solar flux of $\approx 1\,\mathrm{kW/m^2}$ is very convenient for life on earth but (generally speaking) too dilute for effective direct energy conversion. Thus even modest levels of concentration (say 1.5–2) can dramatically increase the temperature at which heat from a solar collector can be extracted efficiently. The economics of photovoltaic conversion can improve significantly with intermediate levels of concentration (say 10–100). At high flux levels (of say $\approx 1,000$) efficient generation of electricity is technically feasible. At the very highest flux levels achievable (say 50,000 or more) certain processes (e.g. laser pumping) become accessible to solar energy utilization which may be of considerable interest. Thus while we may argue over the appropriate flux level for a particular application, there is general agreement that some degree of concentration would be desirable for most applications. Why then are practical solar installations (with some notable exceptions) non-concentrating flat plates? The answer must be that the penalties in cost and complexity associated with optical concentration outweigh its advantages. It is the task of non-imaging optics to mitigate these penalties.

In effect the optics of solar flux concentration has usually been "non-imaging" since the image quality is rarely an issue. But what non-imaging optics has come to signify is the systematic study and development of design techniques that optimize angular acceptance and throughput efficiency for a given flux concentration, a "high collection optics" [1].

3.4.2   *Limits to concentration*

The connection between angular divergence (half angle θ) and concentration (C) can be obtained from Liouville's theorem [2] or thermodynamic arguments [3]. For example, for a two dimensional system (like a parabolic trough) the limit is

$$C = 1/\sin\theta \qquad (3.1)$$

and for a three dimensional system (like a parabolic dish) the corresponding limit is

$$C = 1/\sin^2\theta \qquad (3.2)$$

[If the flux is concentrated at a higher index of refraction (n), these limits are multiplied by n (or n^2) for two (or three) dimensions.] Note these limits apply to reversible systems; fluorescent concentrators beat them by down-shifting the spectrum. It is instructive to see that actually attaining these limits is non-trivial.

Figure 3.4.1 shows a parabolic dish concentrator. From simple geometry one finds $C < 1/(4\sin^2\theta)$!

A similar argument for two-dimensional concentration on a tube gives $C < 1/(\pi\sin\theta)$. The significance of factors like 4 or π cannot be over-stated. For example, it permits a stationary (fixed-mount)

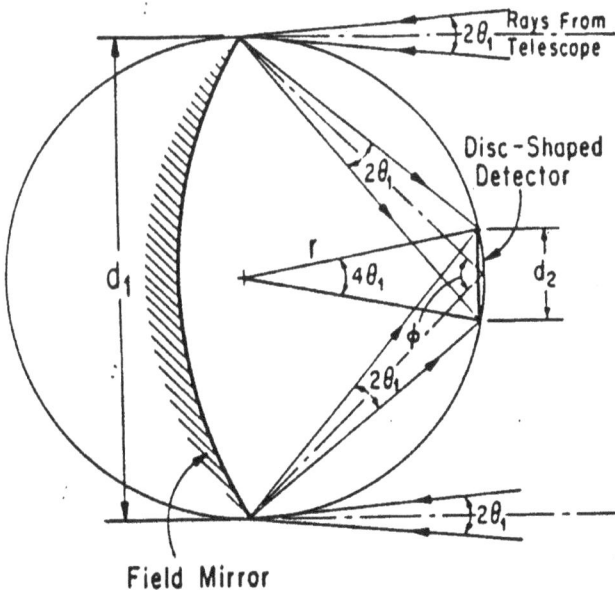

Figure 3.4.1: Concentration limit with parabolic dish

collector ($\theta \approx 35°$) to concentrate rather than de-concentrate! Fortunately, non-imaging optics provides designs which do attain the theoretical limits. More recently these limits have been shown to apply to systems where wave-length dependent effects are important [4] (e.g. holographic optical elements).

3.4.3   *Non-imaging designs*

Two very different design methods have been developed in non-imaging "high collection" optics. The first is the "edge-ray" method where the manifold of extreme input rays become the extreme output rays after reflection (and refraction) in the concentrator. The characteristics of the extreme rays (both input and output) can be arbitrarily specified within wide limits. For example, the output rays may form a focus or caustic at the edge of the exit "aperture" or receiver, or may be parallel to a fixed direction.

This method is illustrated in Fig. 3.4.2 where the extreme input rays forming the wave-front W_1 are refracted by the lens and reflected by the side-wall to become the output wave-front W_2. Along the portion AB of the mirror profile, the rays are focused into the edge of the exit aperture: along the portion BC they are reflected to the wave-front W_2. For both portions, the rays satisfy the condition

$$\int \mathrm{ndl} = \text{constant} \qquad (3.3)$$

In the case shown, the (two-dimensional) concentration achieved is $C = \sin(20°)/\sin(4°)$ because $\theta_1 = 4°$, $\theta_2 = 20°$ are the angles of the two wave-fronts. Take away the lens, let $\theta_2 = 90°$ and you obtain a tilted parabola [5], the original "CPC".

Another instructive example is where the exit "aperture" is a tube. Then the extreme exit rays form a caustic on the tube as shown in Fig. 3.4.3.

A variety of designs for different absorber shapes are illustrated in Fig. 3.4.4.

The other design method relies on a construct called net flux [6] in radiative transfer.

$$\mathbf{J} = \int \mathbf{n} \mathrm{d}\Omega \qquad (3.4)$$

and can be illustrated by the following example:

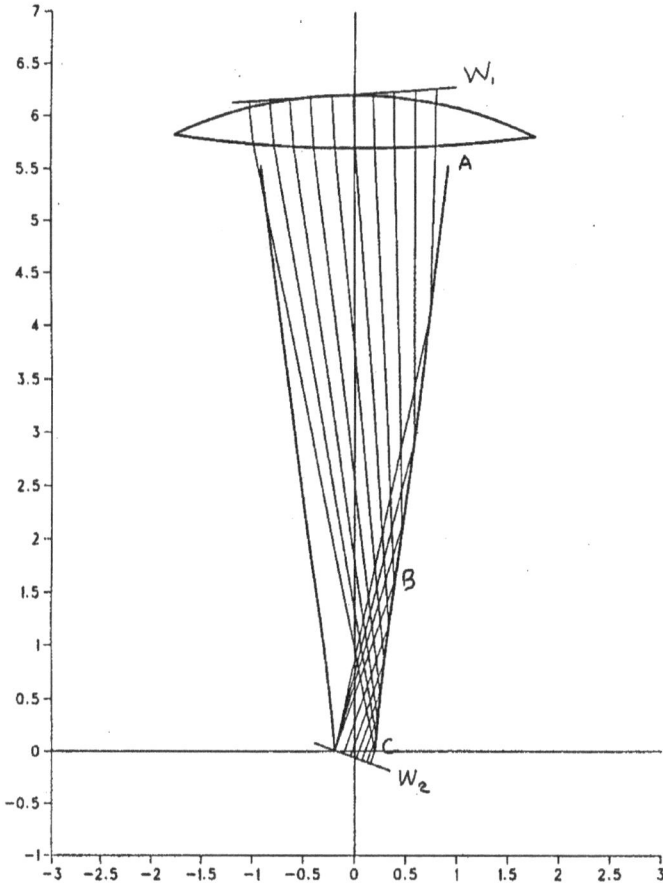

Figure 3.4.2:   Wave-front W_1 is transformed to Wave-front W_2 in this edge-ray design

Take a thermal source such as the aperture of a black-body cavity and connect the <u>flow lines</u> of **J** emanating from the source by a reflecting surface as e.g., the trumpet shaped figure in Fig. 3.4.5.

In certain symmetric situations, like the circular aperture shown, the reflecting surface reconstructs the now virtual source and concentrates it into the smaller real aperture. This design [7] has ideal optical properties in the sense that every ray directed toward the virtual source is concentrated into the real aperture; there are no skew ray losses (as there are for three-dimensional CPC's).

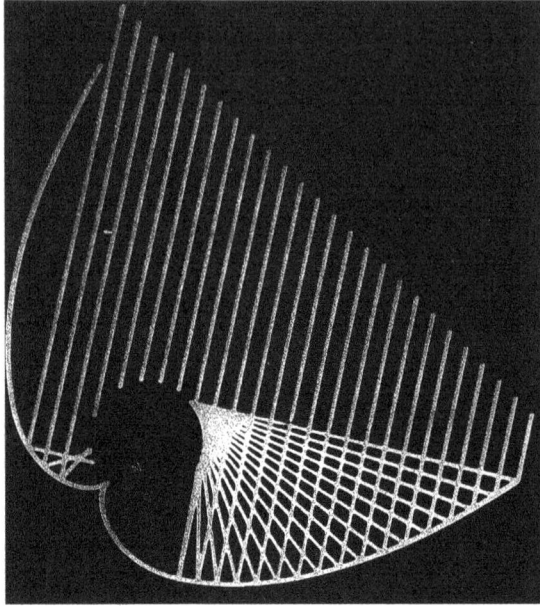

Figure 3.4.3:   Extreme rays are reflected into a caustic on the absorber

Therefore this design, sometimes in combination with lenses may have advantages over the edge-ray designs.

3.4.4   *Selected applications*

3.4.4.1   *Low concentration fixed-mount collector*

It has been recognized for some time that a modest level of flux concentration on a spectrally selective coated, vacuum insulated receiver can give good thermal performance at intermediate temperatures as shown in Fig. 3.4.6.

Since $C < 2(\theta > 30°)$ is compatible with stationary (fixed mount) collection there is considerable incentive to develop low concentration designs. The stationary CPC mirrors with evacuated dewar-type absorbers are used in several commercial collectors (see Fig. 3.4.7).

The concept of integrating the optics with the absorber in a single evacuated tube [8] gives even better performance. Figure 3.4.8

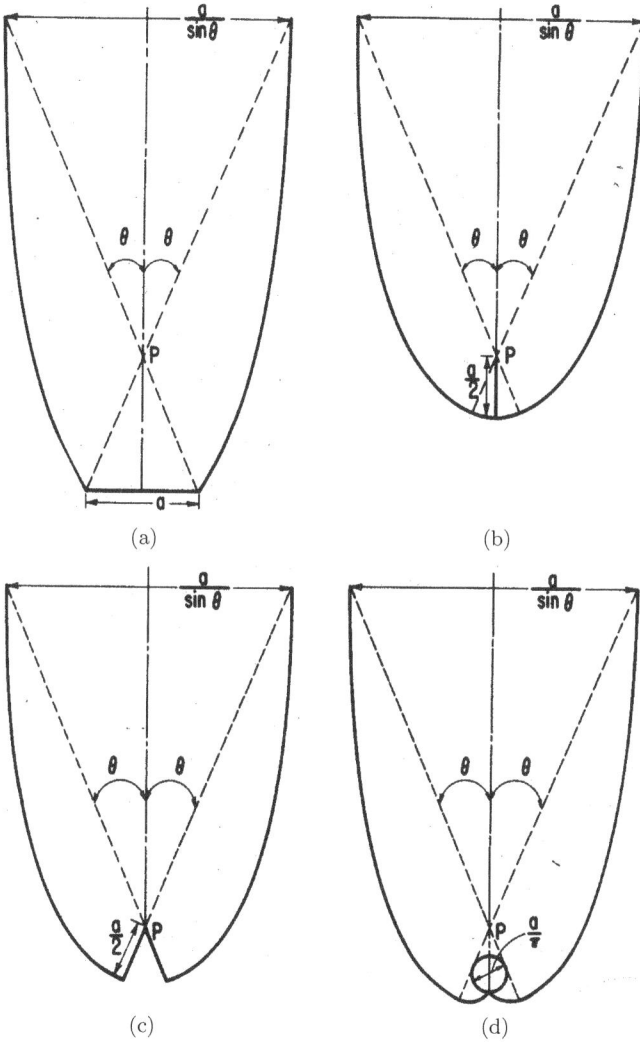

Figure 3.4.4:   A variety of edge-ray designs

shows possible integrated designs while Fig. 3.4.9 shows expected performance.

About 100 prototype tubes based on this concept successfully demonstrated superior performance over more than three years of testing at our laboratory [9]. Our performance goal is to deliver

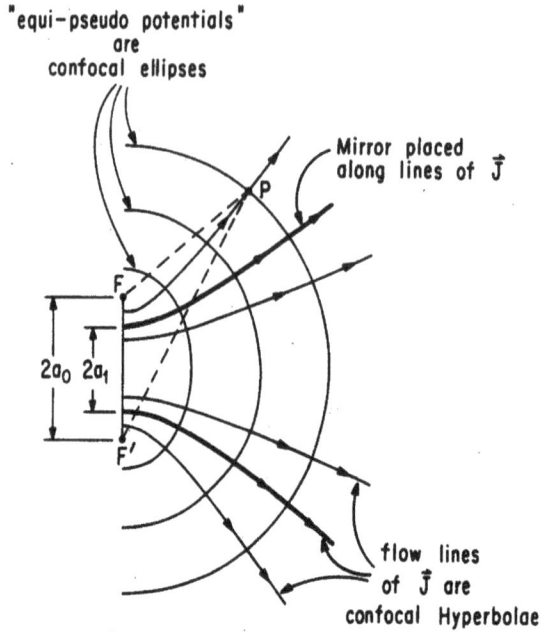

Figure 3.4.5:   Design of the flow-line (trumpet) concentrator

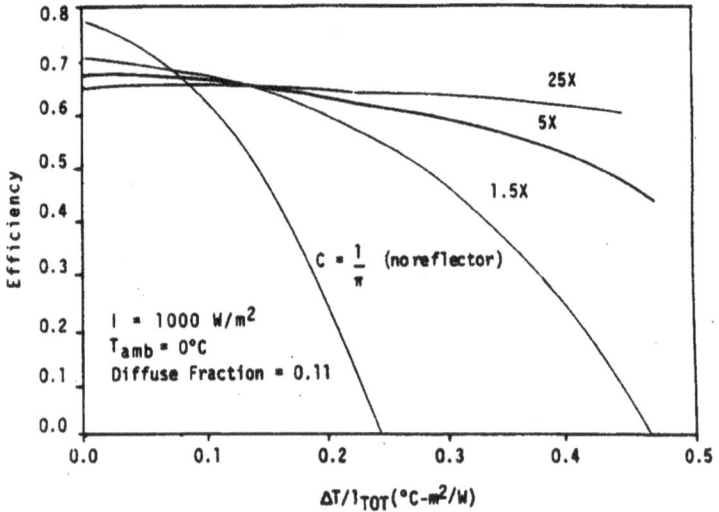

Figure 3.4.6:   Calculated thermal performance curves for evacuated tubular absorbers under increasing levels of concentration. Note that the improvement in increasing the concentration above 5× is marginal

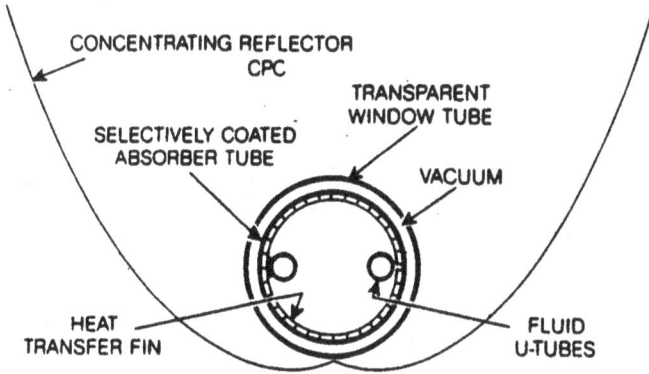

Figure 3.4.7:   Cross-section profile of contemporary evacuated tube CPC according to the basic design developed at Argonne National Laboratories

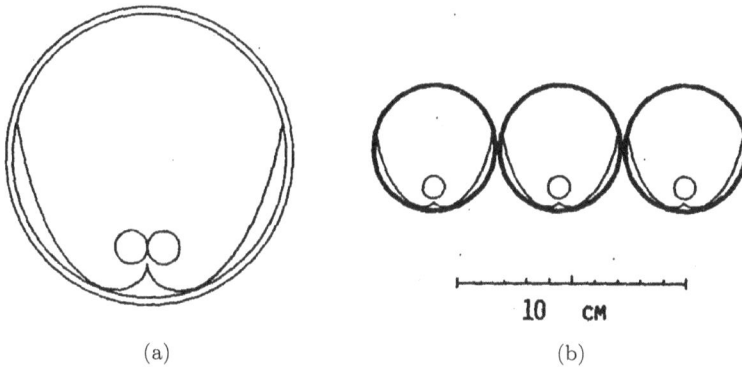

(a)   (b)

Figure 3.4.8:   Alternative integrated CPC designs

thermal energy at a temperature of 175°C with an annual efficiency of 50% while at the same time having a performance and cost which will be competitive with existing collectors at domestic hot water temperatures.

3.4.4.2 *Intermediate concentration for photovoltaic applications*

A non-imaging secondary optical element can be used in conjunction with a conventional focusing lens, either to increase greatly the

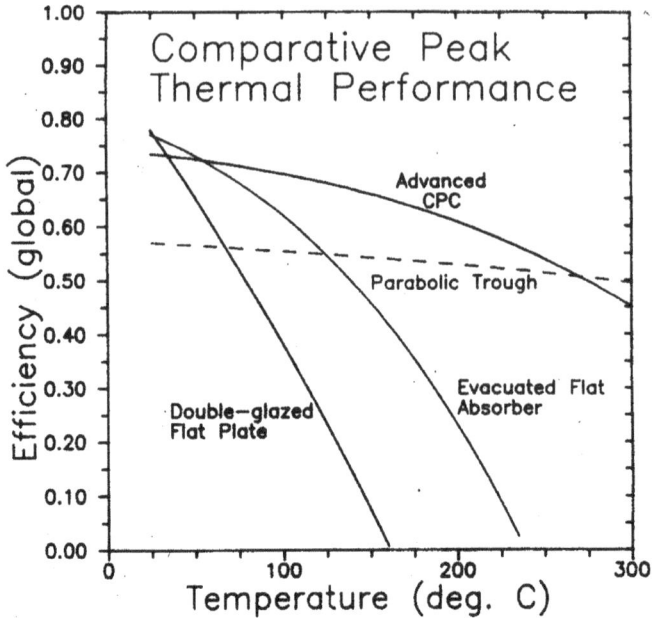

Figure 3.4.9:   Calculated comparative performance of an advanced CPC design compared to collectors that do not employ nonimaging optics

achievable concentration or to relax the required optical tolerances. At the present time, the most developed optical systems are Fresnel lenses. However such systems require accurate tracking and alignment, typically to better than ±0.5° in addition, it is very difficult to achieve concentrations over 500× with a Fresnel lens alone. Two-stage non-imaging concentrators employing Fresnel lenses as the primaries and dielectric totally internally reflecting concentrators as the secondaries offer a means of overcoming these difficulties. A two-stage design can deliver a desired concentration in the high concentration regime (500×–1,000×), and in a more moderate concentration regime (80×–200×), it can provide the maximum possible acceptance angle. In photovoltaic applications these concentrators provide much larger acceptance angles than conventional lens-cell systems with the same geometrical concentrations [10]. It has also been shown that a two-stage concentrator provides a more uniform and consistent

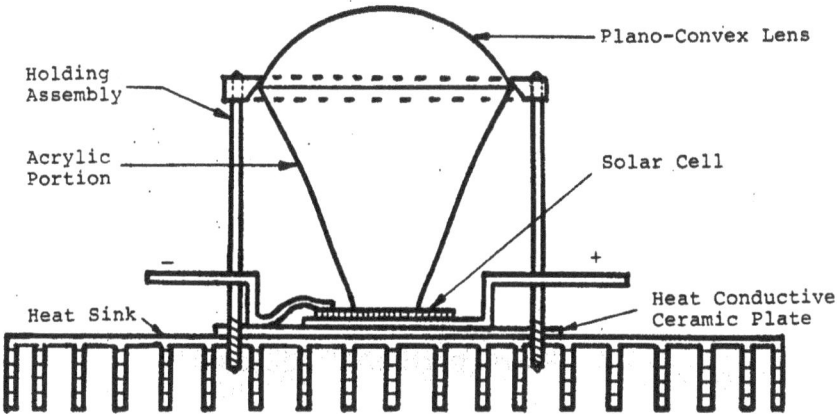

Figure 3.4.10: Dielectric CPC secondary

flux distribution on the cell surface and therefore, the cell's electrical performance can be enhanced.

To demonstrate the technical feasibility and establish the performance of this approach, a prototype of such a two-stage photovoltaic concentrator (4-cell array) was designed, constructed and tested. For this experiment the prototype was designed to achieve a concentration of 144×. The main advantage is seen as its large angular acceptance.

Figure 3.4.10 shows the secondary assembly, while Fig. 3.4.11 shows the dramatic increase in angular acceptance from ±0.5° to nearly ±4°. In addition, we found that the optical losses in the secondary (≈15%) were essentially off-set by improved electrical performance of the more uniformly illuminated cell. Thus, the on-track electrical performance of the lens, or lens-with-secondary, were the same, but the angular tolerance was increased by a factor of nearly 8.

3.4.4.3 *High concentration in point-focus dish solar thermal applications*

Non-imaging secondaries such as the well-known Compound Parabolic Concentrator (CPC), deployed in the focal zone of image forming primary concentrators, offer the optical designer

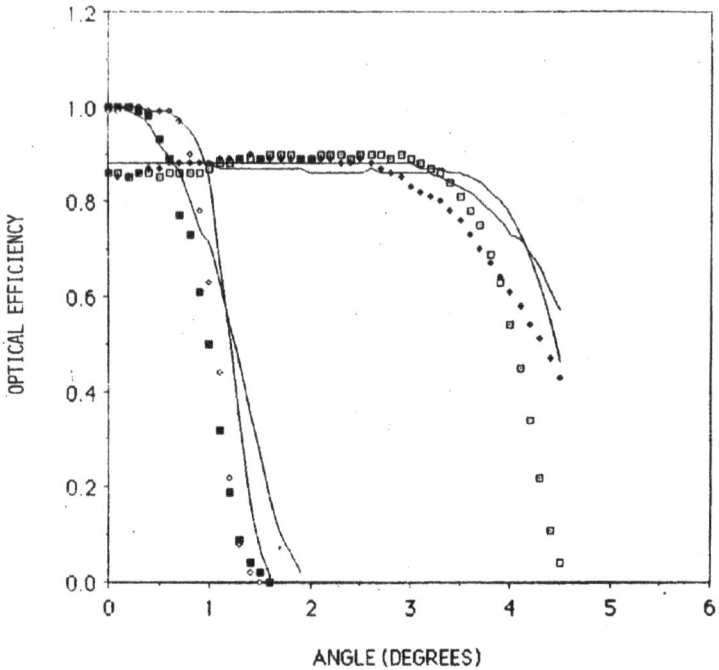

Figure 3.4.11:   Measured angular acceptance of two-stage concentrator. Ray-tracing predictions are shown by solid lines

an additional degree of freedom unavailable with any conventional approach [1]. For a given angular acceptance, such two-stage configurations have the capability of delivering an additional factor of two to four in system concentration. In general this is accomplished without doing anything to the primary. This unique capability can be used either to increase concentration or angular acceptance. For solar thermal applications these parameters are related respectively to increased performance (thermal efficiency) or relaxed optical tolerances (mirror slope error, angular tracking, etc.). Several alternative configurations are possible including the Compound Elliptical Concentrator (CEC), and the flow-line or "trumpet" concentrator in which the reflector is a hyperboloid of revolution. The latter has been developed extensively, both analytically and experimentally. In addition, several versions of a secondary 5× CPC, were fabricated

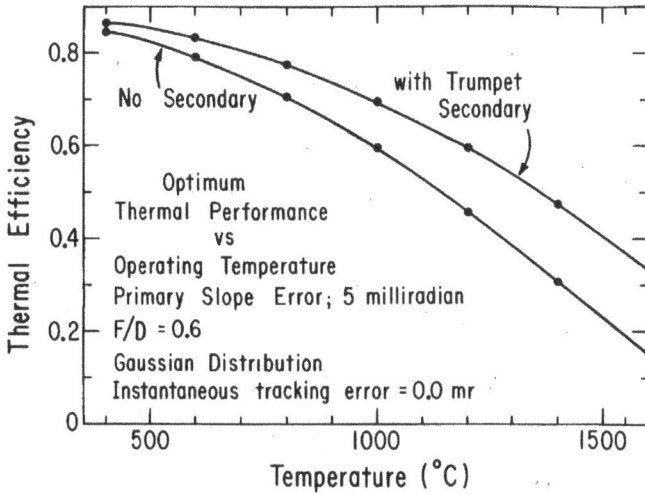

Figure 3.4.12: Calculated performance improvement provided by a nonimaging trumpet secondary as a function of design temperature

and tested at the Advanced Component Test Facility at Georgia Technical Institute [12].

Figure 3.4.12 illustrates the performance improvement provided by a non-imaging trumpet secondary as a function of design temperature while Fig. 3.4.13 shows thermal performance, with and without secondaries in the "high performance limit" of very small slope errors. It is clear that in essentially all regimes of design parameters, the use of a secondary benefits performance.

3.4.4.4 *Ultra-high concentration of solar flux*

As we have emphasized, the concentration limit of terrestrial sunlight is $1/\sin^2\theta_{sun} < 46,000$ in air, and $n^2 \times 46,000$ in a medium with index of refraction $n > 1$. However imaging reflectors achieve less than 25% of this limit. By using a non-imaging secondary concentrator in the focal plane of the primary mirror, the geometrical concentration can be increased by an order of magnitude. For example, the concentration factor for a primary alone is (neglecting shading)

$$C_{primary} \approx \cos^2\phi \sin^2\phi \sin^2\theta_{sun} \qquad (3.5)$$

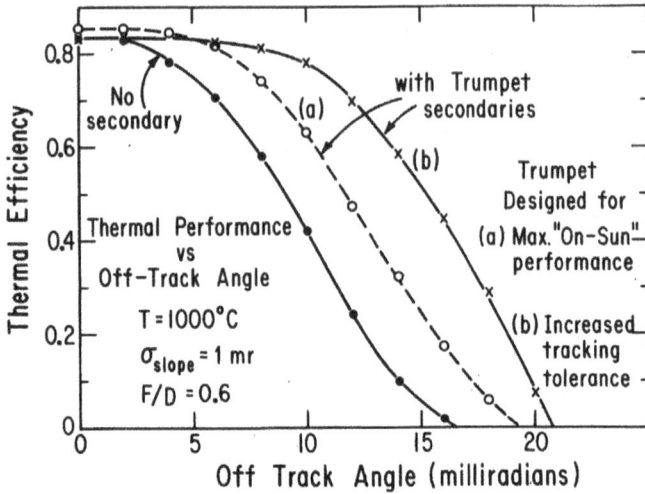

Figure 3.4.13: Calculated thermal performance, with and without secondaries, versus tracking bias error in the "high performance limit" of very small slope error

where ϕ is the rim angle. However, with the addition of a secondary element which is matched to the primary,

$$C_{secondary} \approx n^2 / \sin^2 \phi \qquad (3.6)$$

So that the net concentration of the two-stage system is

$$C \approx n^2 \cos^2 \phi / \sin^2 \theta_{sun} \qquad (3.7)$$

which could, in fact be significantly higher than the limit in air of 46,000 even after various losses are taken into account. We have built a two stage design that has a geometrical concentration of 102,000 by using an 11.5° rim angle primary, and a secondary of index 1.53. With this design one might expect to approach the irradiance that exists on the surface of the sun itself, $7.3\,\mathrm{kW/cm^2}$. With such an intense light source, it may be possible to achieve lasing in materials which efficiently absorb solar radiation.

3.4.5   *Conclusions*

The methods and techniques of non-imaging optics permit us to concentrate solar flux at wide collection angles. This high collection optics can be used to advantage at every level of concentration in order to increase flux, or alternatively to decrease requirements on tracking and on component tolerances and precision.

3.4.6 *Acknowledgements*

This work would not have been possible without the collaboration of my colleagues and students in optics and solar energy; Dr. J. J. O'Gallagher, Prof. W. T. Welford, Dr. W. W. Schertz, P. Gleckman and X. Ning. Our research is supported by the U.S. Department of Energy.

3.4.7 *References*

[1] W.T. Welford and R. Winston, *The Optics of Nonimaging Concentrators*, Academic Press (1978); Chinese edition (1987).

[2] R. Winston, "Light collection within the framework of geometrical optics", *J. Opt. Soc. Am.* 60 (1970) 245–247.

[3] A. Rabl, *Active Solar Collectors and their Applications*, Oxford University Press (1985).

[4] R. Winston and W.T. Welford, "Efficiency of nonimaging concentrators in the physical-optics model", *J. Opt. Soc. Am.* 72 (1982) 1564–1566.

[5] H. Hinterberger and R. Winston, "Efficient light coupler for threshold Cherenkov counters", *Rev. Sci. Instr.* 37 (1966) 1094–1095; V.K. Baranov, "Parabolotoroidal mirrors as elements of solar energy concentrators", *Appl. Solar Energy* 2 (1966) 9–12.

[6] S. Chandrasekhar, *Radiative Transfer*, Dover (1960).

[7] R. Winston and W.T. Welford, "Geometrical vector flux and some new nonimaging concentrators", *J. Opt. Soc. Am.* 69 (1979) 532–536.

[8] J.D. Garrison, "Optimization of a fixed solar thermal collector", *Solar Energy* 23, (1979) 93.

[9] K.A. Snail, J.J. O'Gallagher and R. Winston, "A stationary evacuated collector with integrated concentrator", *Solar Energy* 33 (1984) 441–449.

[10] X. Ning, J.J. O'Gallagher and A. Winston. "Optics of two-stage photovoltaic concentrators with dielectric second stages", *Appl. Opt.* 26 (1987) 1207–1212.

[11] R. Winston and W.T. Welford, "Design of nonimaging concentrators as second stages in tandem with image-forming first-stage concentrators", *Appl. Opt.* 19 (1980) 347–351.

[12] J.J. O'Gallagher, R. Winston, D. Suresh and C. T. Brown, "Design and test of an optimized secondary concentrator with potential cost benefits for solar energy conversion", *Solar Energy* 12 (1987) 217–226.

3.5   Discussion following Roland Winston's Presentation

Goldner: At the beginning of your talk you mentioned luminescent concentrators without any further elaboration. Is there anything inherently wrong with the concept?

Winston: Not that I know about. The concentration of $1/\sin\theta$ gets multiplied by a huge Boltzmann factor. The idea is fine. It is simply a materials problem to execute the idea.

Gordon: In your high flux experiment why did you use a combination of lens and reflector rather than a trumpet-type secondary reflector?

Winston: The lens-trumpet would give more reflections. We were concerned about the possible over-heating of the reflector, and preferred to suffer an extra 3% skew-ray loss than risk having the trumpet melt as it was turned on the sun.

Chapter 4

1991

4.1 Editor's Foreword

Presentations at the 4th Sede Boqer Symposium were heavily dominated by solar-thermal power generation considerations. There was particular emphasis on the problematics associated with two-phase flow: steam and water being the relevant media for the direct steam generation (DSG) system that had been left unfinished at Sede Boqer by the recently defunct Luz Corporation (Fig. 4.1.1). However, in spite of considerable interest in DSG at Plataforma Solar, the European solar test center in Almeria, Spain, its director, Wilfried Grasse chose to present the *PHOEBUS* project, a central receiver solar plant that was being designed for construction and operation in Jordan. Although the project did not subsequently materialize, the importance of Dr. Grasse's paper lies in the amount of detail that went into the system design. The other keynote presentation, by Vern Risser, being a photovoltaic subject, will appear in Volume 2 of this set. However, two additional short presentations from the 1991 symposium have been included in the present chapter. The first was an after-dinner talk by the late Harry Zvi Tabor (1917–2015). The other was a brief paper by the present editor summarizing the costs of solar electricity from the three large-scale technologies that had reached commercialization at that time.

Figure 4.1.1: Unfinished Luz DSG system at Sede Boqer. Zig-zag collector configuration enables gravity to separate water from steam, ensuring that only dry steam exits the northern end of each line of collectors [Photo: D. Faiman]

4.2 *PHOEBUS* — International 30 MW$_e$ Solar Tower Plant (Dr. Wilfried Grasse)

An invited keynote presentation by Dr. Wilfried Grasse (German Aerospace Research Establishment, and Plataforma Solar de Almeria, Spain.)

4.2.1 *Project background*

Starting in 1986, a group of European companies, convinced of the technical and commercial potential of solar thermal central receiver technology, took up an initiative of the Swiss SOTEL group and the German DLR research establishment to commercialize the technology. In 1988, this group was expanded to include new members from Germany and the United States of America, and it was organized as a partnership called the *PHOEBUS* Consortium. In the beginning of 1991 there were nine partners in the Consortium (Table 4.2.1), with

Table 4.2.1:   The *PHOEBUS* consortium partners

Partner	Location
ASINEL	Madrid, Spain
CIEMAT/IER	Madrid, Spain
Didier-Weke[a]	Wiesbaden, Germany
(*Managing Partner*)	Stuttgart, Germany
Fichtner Development Engineering[b]	
Flachglas Solartechnik[b]	Köln, Germany
INITEC	Madrid, Spain
INTERATOM	Bergisch Gladbach, Germany
SOTEL Consortium	Baden, Switzerland
Asea Brown Boveri	
Atlantis Energie	
Bonnard and Gardel	
Colenco	
Centre Suisse d'Electronique	
Compagnie Industrielle Radioelectrique	
Elecktrowatt Ingenieurunternehmung	
Alfred Gudel	
Mini kus, Witta, and Partner	
Gebrüder Sulzer	
U.S. PHOEBUS Associates[b]	
Pacific Gas and Electric	San Ramon, California
Bechtel International	San Francisco, California
(*In an advisory capacity*)	
DLR	Köln, Germany
PSI	Würenlingen, Switzerland

[a]Since 1989, [b]Since 1988.

DLR (Deutsche Forschungsanstalt für Luft-und Raumfahrt, Cologne, FRG) and PSA (Paul Scherrer Institut, Würenlingen, Switzerland) acting in an advisory capacity.

The *PHOEBUS* Consortium was to undertake all activities needed for the detailed design, commissioning, operation, and financing of a plant in order to accomplish the following:

• demonstrate that a solar tower power plant can be operated with continuous high reliability and is able to produce electricity according to demand;

• develop a technical experience base for scaling to commercial size projects;

- demonstrate the potential for larger power plants to provide electric energy at competitive rates.

In two phases (Phase 1A in 1986–87 and Phase 1B in 1988–90), the feasibility study was completed in March 1990. It included predesigns for each major subsystem, in sufficient detail to allow reliable performance analysis and capital cost estimates to be made [1, 2].

4.2.2   *Technical and economic potential of solar tower plants*

Following a DLR initiative, the technical potential of different solar plant concepts (tower, trough and dish) was investigated by using common models and evaluation schemes in order to determine the energetic and economic figures for designed and operating power plants [3]. Annual electrical energy production of all three plant types were compared using the following tools:

- Daily input/output characteristics
- Annual system and subsystem efficiencies
- Diurnal plant performances on typical days

Data from plants designed or already operating were analyzed in detail (Table 4.2.2).

The simulation code SOLERGY, the predictive power of which may be checked as well, was used as the common computational tool. In particular, investigation determined the dependence of plant performance upon the following key parameters:

- Insolation condition: 2000–3000 $kWh/m^2/yr$ (direct, normal)
- Plant size: 30–100 MW_e
- Receiver coolant: air, salt, water/steam, oil, helium

In general, a complete technical-economic comparison consists of the following steps:

- Determination of the annual net electrical energy production at 100 % plant availability
- Availability analysis

Table 4.2.2: Specification of reference plants

Concept	Plant	Absorber coolant/type	Aperture area [m²]	Storage capacity [hr]	Coolant outlet temp. [°C]	Operational date available
tower	10 MW Solar-1	Water-steam/tube	71,000	—	516	1985
	30 MW *PHOEBUS*	Air/volumetric	202,000	3	750	—
	100 MW *PHOEBUS*	Air/volumetric	874,000	6	705	—
	100 MW US-Utility	Molten salt/tube	874,000	6	566	—
trough	30 MW SEGS III-V	Oil/LS2	230,000	—	349	1988
	30 MW SEGS VIIA	Oil/LS3	201,000	3	390	—
	100 MW SEGS VIIIA	Oil LS3	580,000	3	390	—
	100 MW SEGS XIIIB	Water-steam/LS4	700,000	6	395	—
dish	9 kW SBP	Helium/tube	44	—	600	1990
	25 kW MDAC	Hydrogen/tube	87	—	720	1985/86
	30 MW SBP	Helium/tube	126,000	—	620	
	100 MW SBP	Helium/tube	402,000	—	650	

- Estimation of plant cost and calculation of the cost of energy
- Uncertainty analysis of the cost of energy

The three technical concepts considered use the same source of energy, direct normal irradiation, and equivalent steps for converting thermal into electrical energy. However, they differ essentially in the way in which they transform irradiated into thermal energy. This is mainly due to the different geometric "sun-reflector-absorber" configurations and, consequently, the tracking method required to concentrate the incoming sunlight onto the absorber.

Thus, a specific absorber geometry and absorber coolant temperature characterizes each system. In order to study the different technologies in detail, the following types of reference plants were selected:

- Existing and operating experimental plants
- Plants demonstrating the technical concept and feasibility (typically 30 MW_e)
- Commercial plants (typically 100 MW_e)

Furthermore, the following boundary conditions were taken:

- Meteodata for 1976 and 1984 (2850/2370 kWh/m^2 annual direct normal insolation) from Barstow, CA (35° latitude north).
- SOLERGY code for determining annual net electrical energy for all plant variants.
- 89 % average reflectivity for all reflectors.
- Water/steam Rankine cycle (tower, trough) and 9 kW SBP Stirling engine (dish) for power conversion.
- Thermal energy storage (exception: experimental plants and dish/Stirling systems).

As any detailed description of the reflector or absorber technologies would be excessively long for inclusion here, only some basic figures are mentioned. More detailed information can be found in [3].

- Tower details: Glass/metal heliostats ranging from 40 m^2 (SOLAR-1) to 115/150 m^2 (*PHOEBUS*/US-Utility) were used.

The volumetric absorber consisted of a wire mesh cooled by a mixture of cold ambient and recirculated warm air. In the salt variant, liquid salt is heated by about 280K, whereas the SOLAR-1 receiver produces superheated steam directly from the feed-water.

- Trough details: The LS2 and LS3 collector 70 mm-diameter absorber tubes are inside an evacuated glass tube to reduce loss via convection. The selective tube surface coating, which reduces loss from re-radiation, was improved from LS2 to LS3, increasing the oil outlet temperature by about 40K, resulting in a more efficient steam cycle using a re-heater cycle. Whereas the LS2 and LS3 collectors were designed in horizontal loops, potential future LS4 collectors are designed to be inclined by 8°, the northern end being higher to facilitate water/steam separation and to reduce incident cosine losses. (See Fig. 4.1.1). Superheated steam is produced in the saw-tooth shaped LS4 collector set-up. A 12 m wide aperture reflector irradiates about 60 % of the surface of the 140 mm-diameter tube. By contrast, only 45 % of the corresponding LS2 and LS3 tube surfaces are irradiated by reflected sunlight from their troughs, of aperture widths 5.0 and 5.76 m respectively.

- Dish details: A MW-range SBP dish/Stirling plant made up of low voltage clusters is planned. Each cluster will be connected to a main power station via individual mid-voltage lines. The reflector consists of a 1.2 m high, 7.5.m diameter outer ring, over which metallic upper and lower membranes are extended. By plastic deformation of the front membrane, a highly precise paraboloidal shape is attainable, which may be maintained and stabilized during operation by a low partial vacuum.

The various technologies are preferably compared by energy measurements and simulation model calculations. The following types of figures were found suitable:

- daily energy at relevant subsystem interfaces
- monthly system efficiencies
- annual system and subsystem efficiencies
- diurnal plant performance on typical days (clear, cloudy)

For concept comparison, energy produced by the system or subsystem has to be related to a reflector-specific area, which, for physical reasons, is the reflecting system aperture area.

4.2.2.1 *SOLERGY*

The simulation code SOLERGY was used to calculate the energy flows for the system variants considered. Given weather data, plant design data, and subsystem thermal losses and electrical consumptions, SOLERGY determines the net plant electrical output as well as the loss figures at the relevant interfaces. SOLERGY is basically a set of rules on how the energy flow is distributed between the subsystems at any given moment. As input must be given in terms of time and energy requirements, specific plant fluid and thermodynamic process parameters are used for efficiency calculations. Because of the generic nature of this code, it may be applied consistently to systems with a similar power conversion chain.

4.2.2.2 *Input/output characteristics*

In practice, the most important plant characteristic is the correlation between the daily insolation and the daily net electric energy per m^2 aperture area. Such a characteristic is not an efficiency curve, though *daily* efficiencies for certain operational conditions can be taken from it. Since these data are continuously measured in operating plants, they are used, preferentially not only to compare theory with experiment, but also different technologies as well. An example of this is presented in Fig. 4.2.1.

The daily amounts found by measurement and simulation have an approximately linear correlation. Regression line slope angle and intercept are the key parameters influenced by subsystem performances. The predictive capacity of the input/output correlation may be described as follows:

- It is a plant specific characteristic which is nearly independent of the meteorological year considered (see also Fig. 4.2.1).
- Given the annual average insolation in kWh/m^2/d, the corresponding annual output in kWh$_e$/m^2/d may be read from the regression

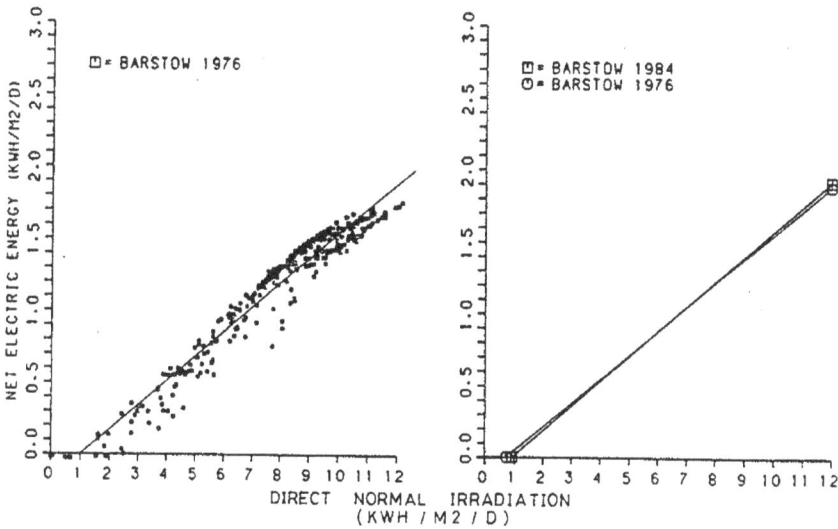

Figure 4.2.1: SOLERGY-simulated daily input/output correlation for 30 MW *PHOEBUS* with Barstow weather data

line. The ratio of both would determine annual plant efficiency at 100 % plant availability.

- Due to regression line interception with the abscissa, plant efficiency increases with annual insolation.

- If the regression lines of two plants intersect, the insolation scale is subdivided into two ranges in which one plant operates more efficiently than the other.

- Since the input/output correlation depends upon geographic latitude, its value has to be noted as a parameter (Barstow: 35° N).

4.2.2.3 *Efficiencies*

Other powerful tools used to comparatively analyze different solar thermal concepts are monthly and annual subsystem efficiencies, which help identify the origin and importance of technical and operational differences, and sensitivity analyses from which the potential allowed for a given system configuration can be estimated. The overall monthly system efficiencies show systematic seasonal tendencies reflecting the reflector/absorber tracking mode. This may be used for scheduling plant maintenance.

4.2.2.4   *Diurnal performance*

Time dependence of plant solar input, interface power and gross output power illustrates the interaction of different plant components. Of particular interest are:

- Receiver thermal output power (MW_e).
- Thermal storage charging/discharging power (MW_e).
- Thermal power of the fossil burner in hybrid operation (MW_{th})
- Gross turbine power output (MW_e)

Daily characteristics may be used to develop optimum operating strategies for typical days (clear, cloudy) and to find the appropriate size for thermal storage according to the load and revenue profile.

Figure 4.2.2 illustrates plant performance calculated by SOLERGY for a clear winter day. If a full turbine load is required for

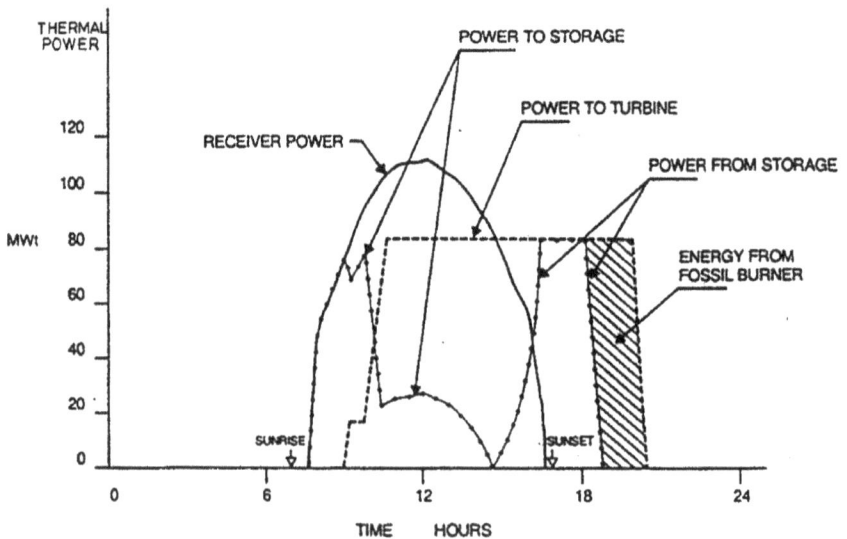

Figure 4.2.2: Typical power utilization on a clear winter day by 30 MW *PHOEBUS* with 3 hr storage and hybrid operation

more than 3 hours after sunset, some fossil energy has to supplement the solar power from storage.

4.2.2.5 *Results*

With a few examples, the results of the comparative analysis are summarized in especially pointing out advantages of *PHOEBUS* type plants versus other solar thermal plant types.

Figure 4.2.3 shows the simulated seasonal variation of monthly average system efficiency for 30 MW plants (using Barstow 1976 meteodata) and indicates the almost constant behavior over the year, whereas trough type plants (no inclination) show substantial performance differences in winter and summer.

Input/output characteristics of the three concepts are shown in Fig. 4.2.4 as calculated by SOLERGY for the meteorological year 1976 at Barstow. On the left, scatter plots and regression lines are depicted for representative 30 MW plants. Whereas the tower and the dish concepts convert the direct irradiation almost linearly into electric energy, SEGS VIIA data points scatter around the regression

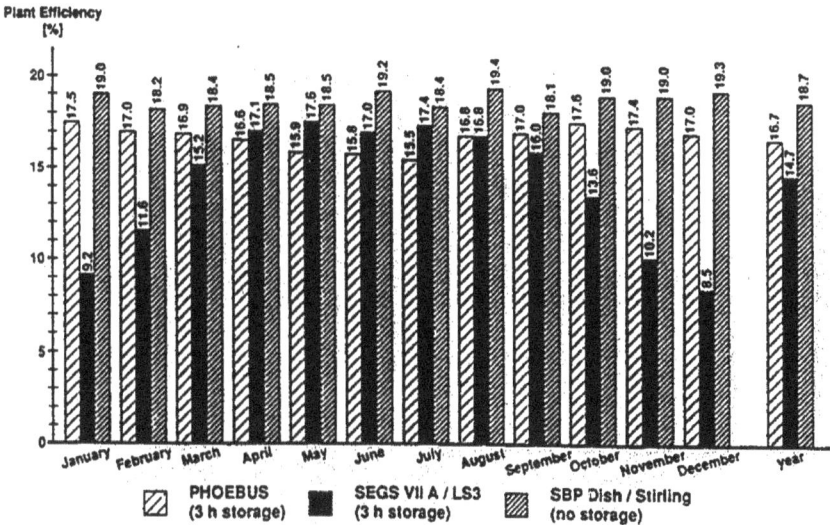

Figure 4.2.3: Simulated seasonal variation of monthly average system efficiency for 30 MWe plants. Meteodata from Barstow, 1976

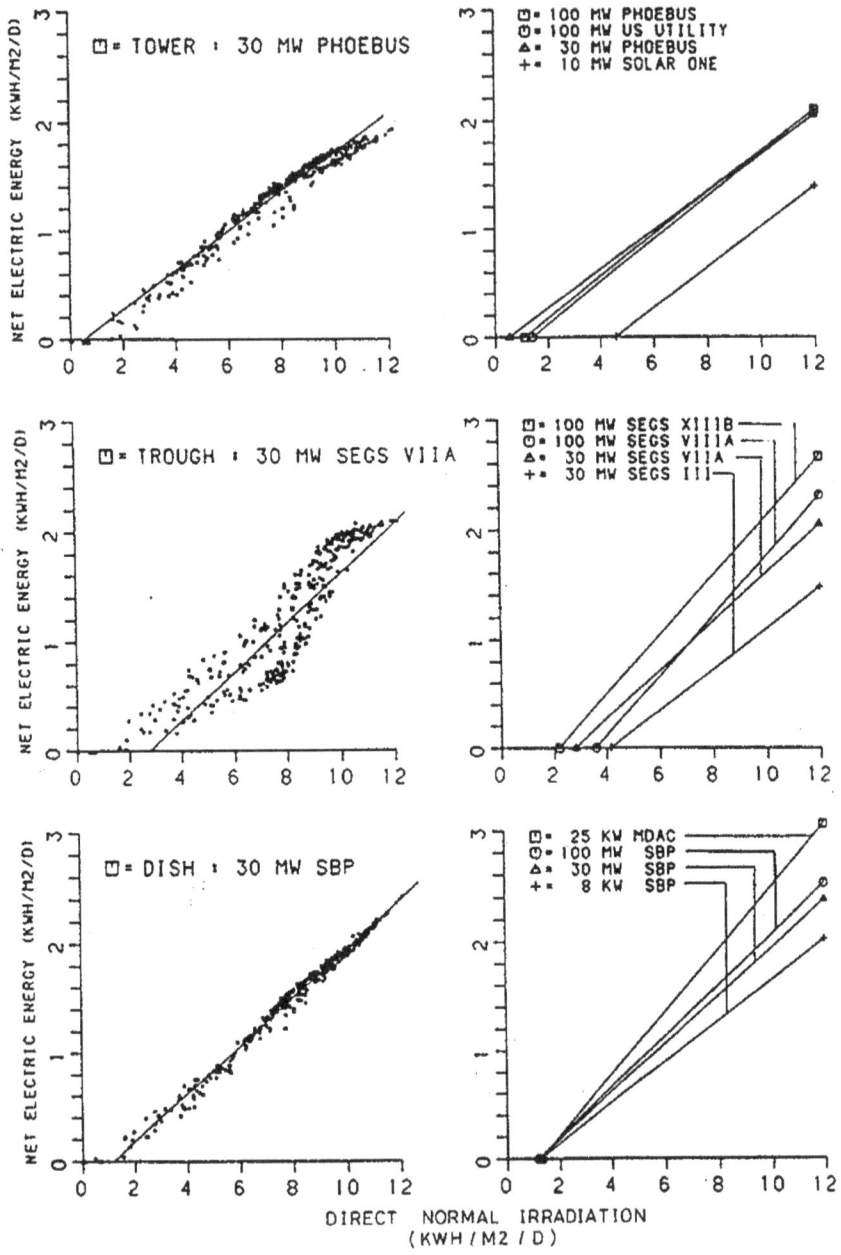

Figure 4.2.4:   Input/output characteristics of solar thermal plants with solar-only operation, from SOLERGY simulations using Barstow 1976 Meteodata

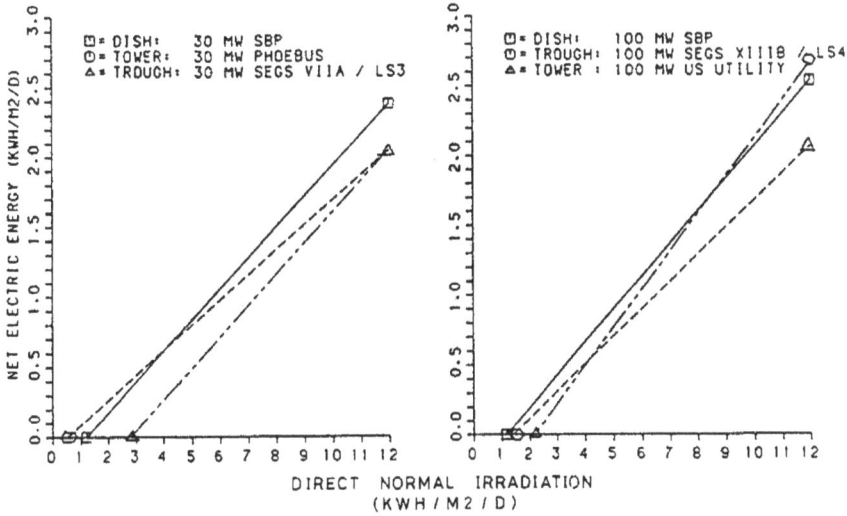

Figure 4.2.5: Input/output characteristics for different solar thermal power plants of 30 and 100 MW$_e$ (SOLERGY simulation for Barstow, 1976)

line. This phenomenon is systematic, since the north-south oriented trough cosine losses are high in the winter and low in the summer. The corresponding irradiation is utilized more efficiently by dish and tower than by the trough.

Comparison by superimposing input/output characteristics for systems with different technologies is presented in Fig. 4.2.5, which shows the dish technology to be the most efficient over almost the whole range of irradiation. This holds true regardless of power range. The tower concept produces electrical power at a somewhat higher efficiency than the LS3, which is the current trough technology. The LS4 technology, as considered for the 100 MW trough plant, is not yet developed.

Annual subsystem efficiencies, a second powerful comparison tool for different solar thermal concepts, are compared in Fig. 4.2.6 (30 MW$_e$). Tower, trough and dish all convert the solar input in a similar manner, but efficiencies of their subsystems differ somewhat. However, in all cases, predominant losses occur at the reflector field, the receiver (absorber) and the turbine, which together determine

Efficiency [%]

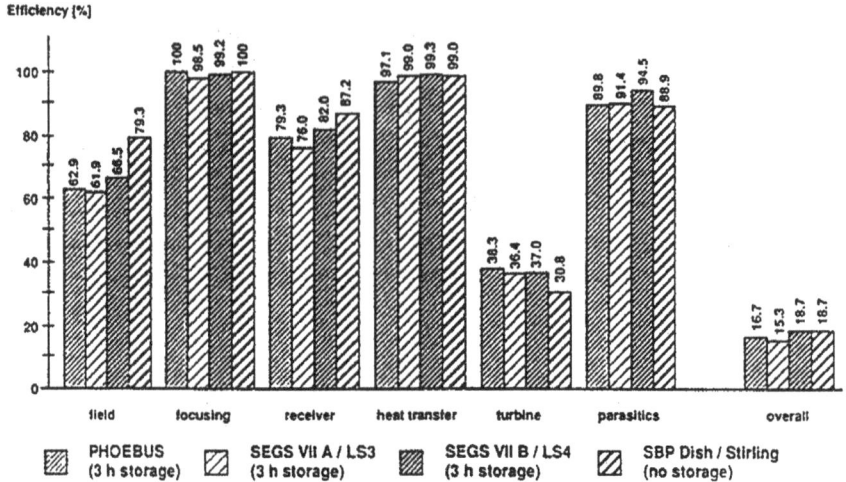

| field | focusing | receiver | heat transfer | turbine | parasitics | overall |

PHOEBUS (3 h storage)   SEGS VII A / LS3 (3 h storage)   SEGS VII B / LS4 (3 h storage)   SBP Dish / Stirling (no storage)

Figure 4.2.6:   Annual subsystem efficiencies of 30 MWe plants

the overwhelming part of the overall system efficiency. In the case of *PHOEBUS*-type plants, there is quite some room for improvement in the existing technology, whereas trough type plants need to develop a new technology (here marked as LS4 with direct steam generation).

A third tool for visualizing solar thermal plant operating characteristics are diagrams showing plant performance on typical days. In Fig. 4.2.7, power per square meter aperture area is plotted for two days in January and June, respectively, at several subsystem interface stations. Again, the trough system operation is significantly different in the summer and winter. Diurnal performance of the solar tower is more balanced over the year and would need less seasonal storage capacity (or fossil back-up heating) in winter, if the nominal performance has to be equal over the full year.

In summary, there are good technical and operational reasons given for developing *PHOEBUS* type plants. This is also the case for their potential economic competiveness compared to other solar thermal options. This is illustrated in Fig. 4.2.8 which was developed on the basis of the analyses given under [3] and [4].

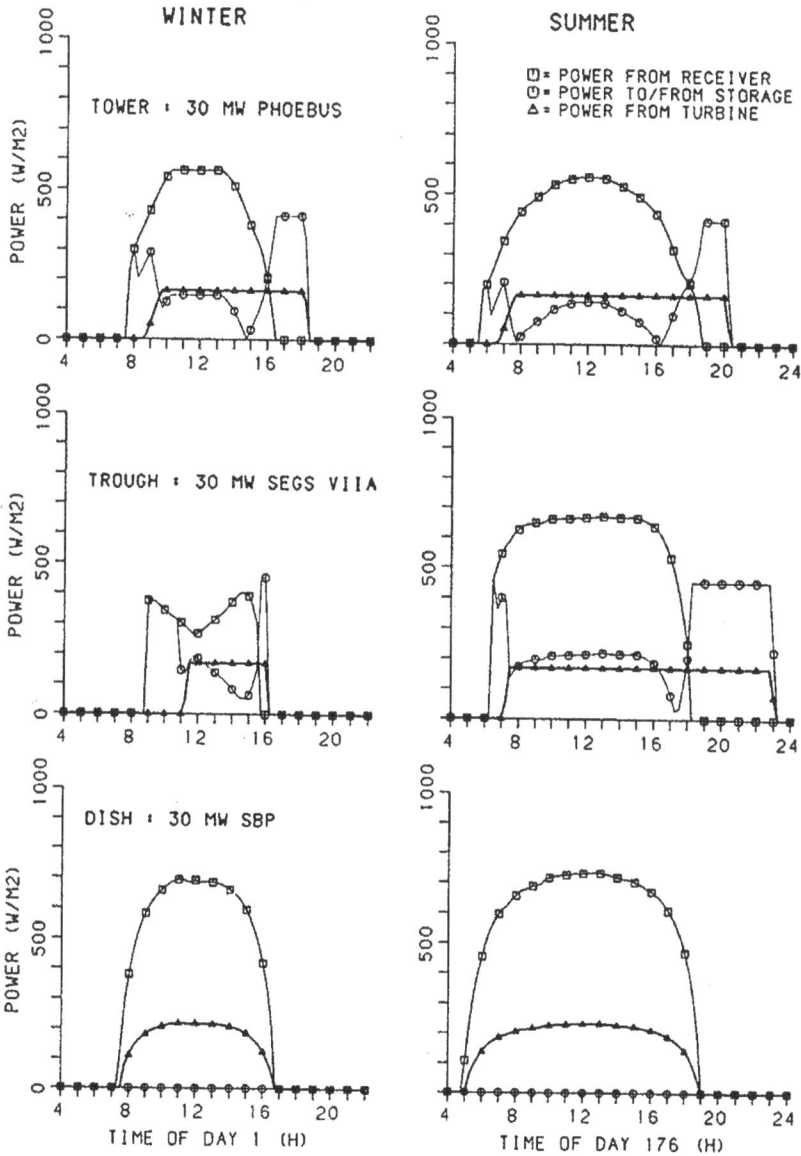

Figure 4.2.7: Diurnal plant performances on selected days (SOLERGY results at several interfaces)

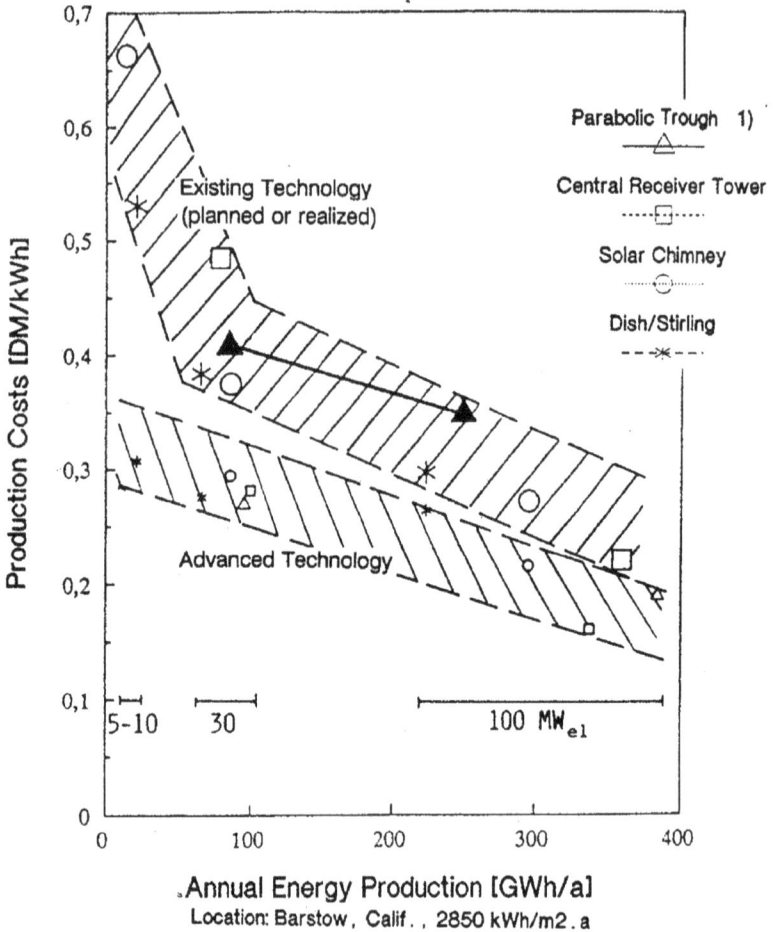

Figure 4.2.8: Production costs of solar thermal power plants (Parabolic trough plant with fossil heater: Today's cost of 0.25 DM/kWh; Interest Rate 7%) [Editor's comment: Average 1994 exchange rate was 1US$ = 1.62 DM]

4.2.3   Site selection

In a study performed in parallel to the *PHOEBUS lA* efforts, DLR with financial support of the German Research Ministry (BMFT) investigated data of 58 countries. The 5 top ranking countries were explored in further detail (i.e. visits for discussions with relevant authorities in the countries) and the Hashemite Kingdom of Jordan was finally selected. In Jordan, eleven candidate sites were investigated with respect to insolation, temperature, rainfall, overcast days, sky clearness, water availability, access, and electric transmission lines. Meteorological measurements started in 1989 at Al-Quwairah near Aqaba (baseline site) and Ras En Naqb and Al Kharana, both near Azraq.

First results of insolation measurements indicate that yearly direct normal insolation in the Wadi Rum area (approximately 50 km north of Aqaba, 800 m above sea level) is equal or even better than data given in the Mohave desert of California where the 25-year average is 2580 kWh/m². Measurements are currently being taken in order to obtain a representative profile of yearly climate data.

4.2.4   *System design*

A series of system optimization studies was conducted in *PHOEBUS* Phase lB to define the optimum plant characteristic for Jordan. They were based on Al-Quwairah site data and California insolation data (Barstow 1976).

Special attention was given to a receiver coolant analysis. The results showed that an air receiver concept was equal or superior to nitrate salt and water/steam concepts in all evaluation categories except its requirement for further technological development. At the end of Phase lA, the decision was made in favor of air. Because of its operational advantages, the air concept is expected to provide the highest annual net electricity yield and, correspondingly, the lowest costs of energy on a remote site in a less industrialized area. Additional advantageous issues are seen in the fact that air, compared to salt, is a non-toxic and non-reactive heat transfer fluid with no need for heat tracing. In comparison to water/steam, thermal storage is easily possible with the well-developed cowper technology used in the steel industry.

To ensure that the plant can reliably contribute to meeting an evening peak demand (as given in Jordan 2 hours after sunset) the plant design includes thermal storage and the capability of burning fossil fuel. This is accomplished with burners located at the air inlet to the steam generator. System design data are given in Fig. 4.2.9 (plant schematic flow diagram) and Tables 4.2.3 and 4.2.4.

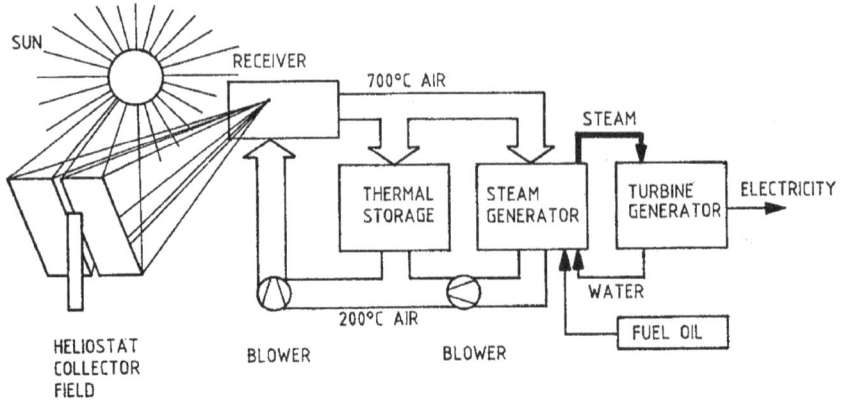

Figure 4.2.9:   *PHOEBUS* plant schematic

Table 4.2.3:   Major plant design parameters

Parameter	Value	Units
Number of Heliostats (115.5 m^2 each)	1751	
Mirror Area	202,200	m^2
Receiver Elevation (Above Grade)	130	m
Receiver Rated Output	101	MW$_{th}$
Incident Power	125	MW$_{th}$
Exit Air Temperature	700	°C
Return Air Temperature	200	°C
Peak Incident Flux	800	kW/m^2
Average Flux	500	kW/m^2
Air Flow	184	kg/s
Receiver Fan Flow	184	kg/s
Pressure Rise	3,000	Pa
Steam Generator Fan Flow	153.5	kg/s
Pressure Rise	5,500	Pa
Duct Dimensions Cold	2.7 × 4.0	m
Hot	3.3 × 5.0	m
Thermal Storage Capacity	250	MWh$_{th}$
Steam Generator Rating	84.1	MW$_{th}$
Air Flow	153.5	kg/s
Live/Reheat Steam Flows	26.4/25.3	kg/s
Temperatures	540/480	°C
Pressures	140/35	bar
Turbine-Generator Gross Output	33	MW$_e$
Condenser Pressure	0.095	bar

Table 4.2.4: Key characteristic plant data

Design/Performance/Operation

Plant configuration	Central receiver
Plant net rating	30 MW$_e$
Annual energy (reference)	~98 GWh$_e$ (Barstow insolation, 1985)
Plant capacity factor	~37% (Barstow insolation, 1985)
Plant solar multiple	1.2
Dispatch strategy	Sun following: fossil firing during 3 hours of evening peak period to start at sunset or at 1 hour before sunset.
Plant design point	Noon, 21 June, 960 W/m^2 direct normal insolation.
Power generation cycle	Rankine steam cycle.
Thermal storage configuration	Thermocline; reference; ceramic bricks (alternative: steel plates).
Tower type	Reinforced concrete, free-standing.
Design life	Overall plant 30 years. Aim for high temperature receiver and gas duct components: 30 years.
Operating staff	21 shifts per week: reduced staff during evening.
Staffing approach	Normal utility approach considering solar-specific requirements. Proposed: 39 men.
Outage times	Scheduled: 14 days in December + 7 single days. Forced: 9 single days in solar-only mode. 0 days in hybrid mode.

Site Parameters

Site location	Southwest Jordan (preliminary reference: Wadi Rum area).
Site access	By road.
Land constraints	None: flat area
Insolation data	Barstow 1985 (reference)
Wind speed at 10 m elevation design point operation	5 m/s or 18 km/hr (to be reviewed considering Jordanian site).
Nominal operation	17 m/s or 60 km/hr.
Heliostat stow initiated	17 m/s.
Survival in high wind stow position (sustained and gusts)	40 m/s or 144 km/hr.
Wind velocity/height dependency	$V = V(10m) \times (H/10m)^{0.15}$

4.2.5   *The heliostat field subsystem*

The collector system, which redirects and concentrates the solar irradiation onto the elevated receiver, is the principal energy supplier for *PHOEBUS*. This capital investment is equivalent to a 30 year supply of thermal energy.

Field layout together with the tower height optimization was performed using the computer code HFLCAL (Heliostat Field Layout Calculation) in combination with the SOLERGY annual performance computer program. HFLCAL (developed by Interatom) optimizes the collector system layout for a given tower height, so that a minimum number of heliostats would supply the requested thermal receiver output. The SOLERGY code (from Sandia) was modified for the purpose of *PHOEBUS*, so that it includes solar-fossil hybrid operating mode. It simulates, on a quasi-steady-state basis, the annual plant performance over a series of 35,040 15-minute increments using real meteorological data (from the Barstow site).

Calculations were made of the annual cost of thermal energy from the receiver for various combinations of tower height, absorber area and field lay-out. The optimum combination included.

- a collector area of 202,000 m^2
- a tower height of 130 m
- a receiver area of ca. 200 m^2(corresponding to a design point concentration factor of ca. 1,000)

Since the absorber peak flux is limited to 800 kW/m^2, a multiple aim point strategy must be applied.

Offers for the collector field were provided from US, German and Spanish suppliers and a design was chosen with an individual heliostat area of 111.5 m^2 at a price of 340 DM/m^2, installed.

Future *PHOEBUS*-type plants may use an advanced, stretched membrane heliostat design. In this concept, two thin metal membranes are tensioned on a hoop, and glass or a second-surface polymer mirror is laminated to the front membrane. A slight vacuum in the plenum between the two membranes gives the front membrane a concave shape. With this design, unit costs for heliostats may decrease to 100–150 DM/m^2.

4.2.6 *The receiver subsystem*

The receiver absorbs the redirected solar flux from the collector system and transfers the heat to an air stream. An innovative receiver (called volumetric type) has been designed [5] in which the basic absorber element is a knitted mesh of fine metal wire. The mesh is mounted in a loose coil such that the concentrated flux is absorbed gradually as it penetrates the coil. Atmospheric pressure air is drawn through a total of 600×7 absorber modules forming a receiver of elliptical shape with $17\,\text{m} \times 18\,\text{m}$ and tilted 15 degrees towards the heliostat field (Fig. 4.2.10).

Because the flux varies over the absorber area, the air flow must be adjusted proportionally; this is achieved by an orifice plate arranged behind each module. A selected pressure drop of approximately 300 Pa serves to stabilize the airflow under the influence of wind.

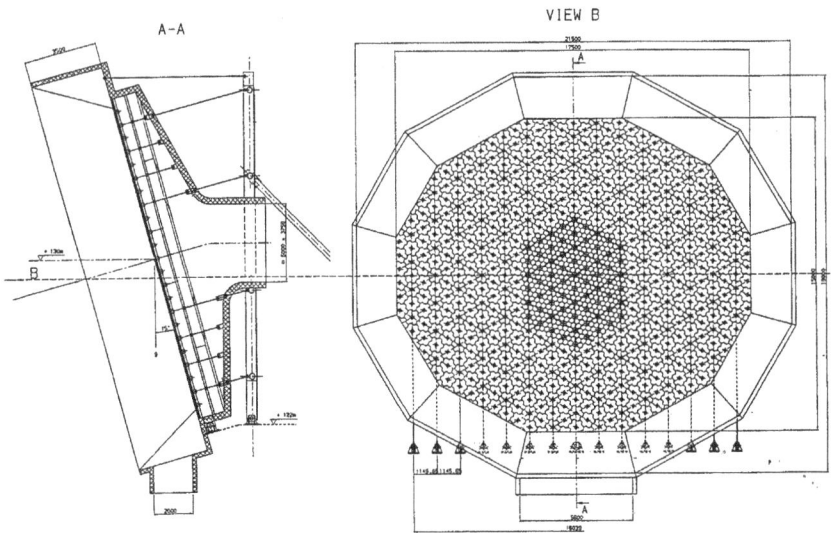

Figure 4.2.10:   *PHOEBUS* air receiver configuration

The modularization of the absorber has several advantages over tube receiver designs, including:

- Large-volume production of identical units
- Small-scale testing of units
- Uninterrupted operation in case of failure of individual modules
- Easy replacement in case of damage

Surrounding the absorber is an ellipsoidal cone which performs three functions. First, it acts as a secondary concentrator to reflect onto the absorber that portion of the incident flux which would normally fall outside the periphery of the absorber. Second, it acts as a re-injection manifold for the warm air returning from the steam generator and thermal storage. Third, the cone supports at its front periphery a radial in flow injection nozzle for an air curtain system. All those means help

- minimizing wind influences;
- providing a uniform incident flux profile on the absorber;
- increasing the portion of recirculated air (from 60% to potentially 90%).

Experimental data exist from small-scale experiments (on the Plataforma Solar, in 1988 and 1989). A more representative system experiment comprising a receiver and a thermal storage unit is under preparation for an energy input of $2.5\,\mathrm{MW_{th}}$, again to be performed on the Plataforma Solar in Almeria, Spain.

4.2.7   *The thermal storage subsystem*

The *PHOEBUS* plant was designed with receiver output exceeding the steam generator peak requirement by 20% (solar multiple of 1.2). A corresponding sizing of the storage system would lead to a storage capacity of 1.5 hours full-load operation. However, in order to have the possibility to demonstrate complete coverage of the Jordanian peak demand period in the evening hours, a storage capacity of 3 hours of full load was chosen in the preliminary design.

Design of the *PHOEBUS* storage system follows rather conventional technologies as used in the steel industry. The storage medium is a refractory ceramic material (with 60% of Al_2O_3) formed in the shape of hexagonal bricks with vertical air passages.

During a charge cycle, hot air at 700°C enters the top of the vessel and exits at the bottom at 200°C. During a discharge cycle, the flow is reversed with air exiting at somewhat less than 700°C. The conceptual design required a useable storage capacity of 250 MWh_{th}. It contains a storage material mass of 7,400 metric tons, and was offered for a price corresponding to 50 DM/kWh capacity.

4.2.8 *The steam generator system*

The design chosen by the *PHOEBUS* consortium is similar to heat recovery steam generators used in combined cycle power plants. The steam generator system consists of the following:

- Drum type, forced recirculation 84 MW_{th} steam generator with separate superheater, reheater, evaporator and economizer sections.
- Duct burner, capable of firing heavy and light fuel oils.
- Fuel oil storage and delivery system.

The steam generator heat source is atmospheric pressure air or burner exhaust gases or a mixture of the two, at 700°C. A forced recirculation design was selected to minimize water treatment requirements and simplify plant daily startup procedures. The duct burners have been designed with a turn-down ration of 8:1, thus it is possible to maintain steam generator output over a wide range of receiver and thermal storage operating conditions.

4.2.9 *The electric power generation subsystem (EPGS)*

The EPGS of *PHOEBUS* is based on the high-efficiency reheat turbine cycles currently in use in the latest solar 30 MW_e SEGS plants in California, which use parabolic troughs as collectors.

The turbine-generator is a double case, single flow machine with a speed reduction gearbox between the high-pressure and low-pressure turbines. The system also includes the surface condenser, the feed water system (heaters and pump) and instrumentation and controls.

The cooling water system uses dry cooling towers to essentially eliminate plant water consumption. To minimize the effect of high ambient temperatures, the cooling towers operate only at night when air temperatures are lowest. Cooling water is stored in large neoprene tanks for use during the day.

The turbine is designed for rapid daily startup and shutdown, and low thermal stresses during transient operation. Gross cycle output is $33\,MW_e$ with a condenser pressure of 0.1 bar. The corresponding gross cycle efficiency is 39.4%.

4.2.10   *Remaining subsystems*

In addition to the subsystems already mentioned, preliminary design of the following subsystems/components was accomplished during Phase 1B efforts:

- Heat transport system with hot and cold air ducts, fans, and shut-off dampers.
- All components are rather conventional with the exception of the hot air duct where the insulation is placed on the inside in order to avoid problems associated with thermal expansion due to air temperatures of $700°C$.
- Instrumentation and control for supervision and overall operation of the plant. It includes the solar energy management system, a microcomputer which determines, in advance, the daily program of plant operating modes. It also includes a thermodynamic model of the plant.

4.2.11   *Overall performances*

The SOLERGY code as amended by the *PHOEBUS* Consortium (see above) permits drawing plant characteristics as given in Fig. 4.2.1 for the cases solar-only and solar/fossil hybrid operation, respectively. Such plant characteristics correlate daily net electric output with daily irradiated normal incident power. Together with a site-specific and statistically relevant irradiation diagram, a sufficiently reliable estimate of the maximum possible yearly output can be derived from such characteristics. Irradiation profiles for the Jordanian site are being measured at present.

For the design as defined in Phase 1B, a yearly net electric energy output of 98 GWh was determined for the solar-only operational mode.

4.2.12 *Implementation and next steps*

The *PHOEBUS* consortium foresees a period of ca. 3 years to bring the 30 MWe *PHOEBUS* plant into operation on a Jordanian site.

In a preceding Phase 2A, called the project development, techno-logical development of the volumetric air receiver concept (see above) must be completed and the organizational and financial pre-requisites established.

4.2.13 *Funding requirements*

An investment cost estimate has been made for a first-ofits-kind *PHOEBUS* Plant, which includes all project costs from the start of project development activities to handing over the plant for commercial operation. The cost calculation is broken down into 20 categories, as shown in Table 4.2.5. The direct cost portion (items 1–12) is based on 1989 offers and prices and totals ca. 228 MDM. Engineering and project development costs (items 14–17) require an additional 42 MDM. The overall plant investment, escalated to 1994 (items 18–20), is 330 MDM. The annual operation and maintenance (O&M) cost includes all expenses for labor, materials, and fossil fuels. With a plant operation of 3 shifts per day and 7 days per week, the annual O&M cost, escalated to 1994, amounts to 6 MDM.

A preliminary funding strategy was developed during Phase 1B which needs to be verified in the forthcoming project development phase. It foresees the following contribution categories:

Industrial contributions: 56%

- In the form of equities from participating companies, the Jordanian partner and Islamic development funds (32%).
- In the form of loans taken from various banks and funds (24%).

Table 4.2.5:   Investment cost estimate and operation and maintenance cost estimate [Editor's comment: Average 1994 exchange rate was 1US\$ = 1.62 DM].

Investment	Cost estimate ($\times$ 1000 DM)
1. Structures and improvements	10,000
2. Tower	9,400
3. Collector System	70,106
4. Receiver System	10,840
5. Heat Transfer System	10,824
6. Thermal Storage System	47,500
7. Steam Generation System	12,460
8. Steam Turbine-Generator	19,200
9. Electric Power Generation Systems	8,844
10. Balance of Plant	12,180
11. Instrumentation and Control System	5,892
12. Electric Equipment	10,330
13. Contingency	22,758
Total Direct Cost	250,334
14. Engineering and Construction Management	15,000
15. Site and Infrastructure (Owner's Costs)	16,800
16. Project Development Costs	4,100
17. Engineering and Construction Insurance	5,770
Total Capital Cost	292,004
18. Escalation (1989 to 1994)	31,500
19. Interest During Construction	2,135
20. Financing Fees	4,122
Total Investment cost	329,761
Operation and maintenance	
Operating Personnel	863
Maintenance Materials	1,490
Overhead	530
Fuel Oil	1,300
Insurance	990
Total	5,164
Escalation (1989 to 1994)	956
1st Year Annual Cost	6,120

Grants from local or federal governments and the European Community: 44%.

Based upon this financial plan the 30 year cash flow analysis predicts a required power sales price of 11.2 Dpf/kWh dynamic unit cost in 1989 prices over the lifetime of the plant (30 years). This is considered to be close to "avoided costs" in Jordan. Later *PHOEBUS* plants of higher power rating will, without public support, deliver electricity at not much higher costs than is expected for the first of its kind *PHOEBUS* plant. [Editor's note: In 1989, the average exchange rate was 1 US\$ = 1.88 DM, and 1DM = 100 Dpf.]

4.2.14 *References*

[1] International 30 MW$_e$ Solar Tower Plant, Feasibility Study Phase lA, Presentation of Results, Ministerio de Industria y Energia, Madrid/Spain (December 1987).

[2] *PHOEBUS*, a 30 MW$_e$ Solar Tower Power Plant for Jordan, Phase lB-Feasibility Study-Volume 1: Main Results, Stuttgart/ Germany (March 1990).

[3] M. Kiera, W. Meinecke, H. Klaiss, Energetic Comparison of Solar Thermal Power Plants, in: Proceedings of the 5th Symposium on Solar High Temperature Technologies, Davos/CH (1990).

[4] P. Klimas, M. Becker, Status of the Second Generation of Central Receiver Technologies, in: Proceedings of the 5th Symposium on Solar High Temperature Technologies, Davos CH (1990).

[5] H.W. Fricker, Tests with a Small Volumetric Wire Receiver, in: Proceedings of the 3rd International Workshop on Solar Thermal Central Receiver Systems, Springer Verlag (1986).

4.3 Discussion Following Wilfried Grasse's Presentation

Weiner: What was the source of the insolation data that went into your evaluation of a suitable site for *PHOEBUS*?

Grasse: Institutions in each of the candidate countries were asked to furnish local data. It is of course a widely recognized problem that radiation data statistics need to be seriously improved around the world.

Faiman: I should just like to comment that if pyrheliometers were used to measure the original data then the results would tend to be conservative since a dirty or misaligned instrument could only give an under-reading (assuming that it was correctly calibrated).

Roy: I was surprised at the heliostat efficiency figures of 61%, 64%, etc., you mentioned. Would you please be more specific?

Grasse: These figures are our design goals. Present experience indicates optical efficiencies well in excess of 50%. Our figure includes reflectivity, dirt, availability, tracking errors, etc. With all these ingredients we believe that efficiencies in excess of 60% are feasible. Luz, for example, reported such figures for their mirrors.

Faiman: What quantities of washing water will be needed in order to maintain your projected power output under the conditions that *PHOEBUS* will encounter in the Jordanian desert?

Grasse: I cannot give you precise figures but, based upon our Almeria experience and figures from California published by Luz, I believe the quantity will be small.

Roy: What was included in the price quotations you received for the heliostats?

Grasse: Everything (including, for example, all cabling and computers) except for the civil engineering work necessary at the site. Delivery costs were also not included because the final site was still unknown when the bids were solicited.

Question: To what extent was the receiver for *PHOEBUS* based upon experience you had already gained at Almeria, and what were the chief factors that needed to be taken into consideration in scaling up from 200 kW to 2 MW?

Grasse: The principle behind the two receivers is similar but, as you noted, the scales are different and different engineering firms are involved in the execution of these designs. Our principal concerns in the scale-up related to: wind stress effects; the percentage of air that will need to be recirculated; and the physical dimensions of several components.

Kashin: Does the receiver design include secondary concentration?

Grasse: In a relatively unsophisticated way, yes: Plane mirrors are used to redirect some of the spilt radiation.

Epstein: I noticed that the air flow rate in the receiver is higher than that at the inlet to the steam generator. What has happened to the balance?

Grasse: Part of the air is recirculated to the receiver through the storage loop. So it should not necessarily equal the volumetric flow through the steam generator.

Tabor: Heat exchangers tend to be a weak link in the design of efficient solar energy conversion systems. Could you add something about the design of the air-steam heat exchanger that *PHOEBUS* will employ?

Grasse: The steam generator, of a forced drum type, was designed by the Sulzer Company to perform according to the specifications listed in Table 4.2.3. The design was then checked by Bechtel Corp. Both of these companies are considered specialists in the field so we believe this sub-system will perform satisfactorily.

Tabor: There is a large drop in thermodynamic potential in going from $750°C$ at the receiver outlet to $540°C$ in the steam generator.

Grasse: This is indeed a penalty that we shall have to pay with *PHOEBUS-1*. But feasibility studies are underway to enable later *PHOEBUS* plants to operate in a combined gas-cycle/steam-cycle mode so as to reduce exergy losses.

Yanir: How will *PHOEBUS* compare with the performance of photovoltaic plants?

Grasse: Our aim is not to compete with photovoltaic plants. At present, on the large scale, solar-thermal is much more cost-effective than photovoltaics — mainly thanks to the achievements of Luz. Photovoltaics are, at the present time, only suitable for small-scale application.

4.4   Some Concluding Remarks (Dr. Harry Tabor)

An after-dinner talk by Dr. Harry Zvi Tabor, (The Scientific Research Foundation, Jerusalem, Israel)

4.4.1   *Introduction*

We here are primarily interested in solar-energy as a possible substitute for fossil fuels. This series of annual symposia — for which we should all be very grateful to our Sede Boqer colleagues — has been limited to solar electricity primarily because it is a most versatile form of energy: it can easily be transported over long distances (even longer if the technologies of super-conducting cables and hydrogen storage will be adequately developed) and, by injection into a national grid, can substitute for fossil energy on a large scale

The range of magnitudes is very wide — of the order of 10^{15}: 1, i.e. from microwatts for a solar watch to gigawatts for a power-station, which allows for a range of technologies for harnessing the sun, some of which are more convenient than others. As R&D continues, the 'best' technologies move along the range. Thirty years ago, a 1 kW solar pump would almost certainly have been a thermodynamic conversion unit (or a windmill, where the wind regime was suitable): today, at this size, PV has taken over. At what upper size is thermodynamic conversion preferable to PV is a subject for debate — sometimes quite heated debate — but size is not the only criterion. Application and location, i.e. local maintenance capability, play a part.

Almost everyone in this room is interested in the top end of the range, i.e. solar electricity for large installations. Even those PV developers, who speak kindly about PV for under-developed villages, have their eyes on the power-station field, for the obvious reasons that this would allow them to enter expensive sophisticated production processes, and it is a lot easier to sell one 1GW power-station than 500,000 2kW units for under-developed villages!

At the bottom end of the scale, where PV stands supreme, we should realize that we are not really talking about selling energy but convenience. (Indeed, as many of you know, at one time the

PV cell was a negative producer of energy, more being used in its production and distribution than it could generate in its lifetime). In the final analysis, until PV reaches a viable MW stage, its total contribution to the energy supplies of the world will be negligible. But it is convenient — and for the people involved — an exciting area — and it can satisfy some specific needs.

In the intermediate-size range — say less than 10 kW — PV can be very useful. One illustrative example is the World Health Organization's EPI. (Extended Program of Immunization) Cold Chain — program for supplying and maintaining a chain of vaccine centers. This requires refrigeration in areas often without electricity supplies and kerosene-operated refrigerators have proven unsatisfactory because of poor temperature control. The WHO has recognized the potential of PV-operated refrigerators and now has more than 1000 such units dotted over the African continent. Note that, for this application, energy-storage (i.e. a large battery) — or some form of backup — (e.g. a small diesel generator), is essential. This organization deserves much credit for having realized that of primary importance in the developing world for the harnessing of solar energy, is the training of local technicians for the installation and maintenance of the units, i.e. the refrigerators, the PV panels, controls and batteries — and they have produced a series of handbooks just for that purpose.

Another area is in the utility field i.e. for supplying householders in areas where the grid is not too close. Whilst Dr. Risser earlier today discussed grid-connected PV systems, the Californian power company, PG&E, has conducted a number of studies illustrating cases in which the supply of PV units to individual houses works out cheaper than electricity from a grid — when the cost of the connection is taken into account (and our Ministry of Energy has been studying this in the Klil Village experiment). A fundamental difficulty for individual PV units is the absence of statistical-load-sharing. The peak power taken by a householder can be 50x the average power use — which no PV unit can handle unless it includes very large batteries with high-discharge capability, whereas, when connected to a utility, he 'borrows' from the net and, only if many people want maximum power at the same time, is there trouble (a so-called 'brownout'). Indeed, some utilities offer a 'bribe' to selected customers if they will agree to accept cuts at peak hours (so that the installed capacity can be kept low).

4.4.2   *Large-scale electricity production*

4.4.2.1   *Solar-thermal electricity*

The large-scale production of electricity from the sun being the main interest for this seminar, it is ironic that the Luz Corporation is not represented here. No group in the world, either of private entrepreneurs or a national — or even international — group has made such a vital and dramatic contribution to large-scale solar electricity. They developed excellent technology. (I have, on previous occasions, referred to the comparison between the Luz plant and the Shuman $50kW_e$ plant built in Egypt in 1913, which also used horizontal parabolic tracking trough concentrators with the difference that the Luz plants yield 5x the electricity per m^2 of the Shuman plant). Their second critical contribution was that they broke the vicious circle — that no utility would risk its money (US$100 million) on a plant, the like of which had never been built before, but that without utility approval, no such plant would ever be built. They introduced the concept of 3rd-party financing — that required enormous courage on their part (a) to promise to deliver a given amount of electricity when requested whilst not asking for money but only a commitment to buy the electricity at an agreed price: (b) on the basis of their estimates of what the plant would cost plus the subsidies of the Federal and California State governments, they could make a package that would look reasonable to private investors — and then go out and find the investors — a hundred million dollars at a time! What is further astonishing is the short time-span. Before the first Luz plant in California, completed in 1985, no one really believed in multi-MW solar plants as economically justified, even if they could be shown to be technically viable — which itself was questionable. The $10MW_e$ central receiver (power tower) in Barstow, California, (completed in 1982) was a 'beautiful' exercise. In the words of one of the people responsible for the *Eurelios* $750kW_e$

central receiver plant in Italy: "it is as beautiful as it is expensive!" After all the data from the Barstow plant had been recorded, it was dismantled, as it did not pay for its maintenance, let alone the capital cost. Thus, a seminar such as this one, had it been held in 1984, could only have presented 'paper schemes', largely influenced by wishful thinking.

The solar-pond held some promise but was clearly limited to very specific areas. For multi-MW$_e$ outputs, wind machines, in windy regimes, offered a better option.

This is not the place to discuss the financial difficulties of the Luz company, but these were *not* the result of the Luz system being technically or economically non-viable but, as explained in the lecture of Mr. Lotker (of Luz) [1] at the recent ISES meeting in Denver, a direct consequence of US conventional energy sources receiving subsidies whilst the US DOE withdrew them from the renewable-energy program. I show here a diagram that, to me, summarizes what may be called "The Luz Enigma" i.e. how a highly successful company can go under.

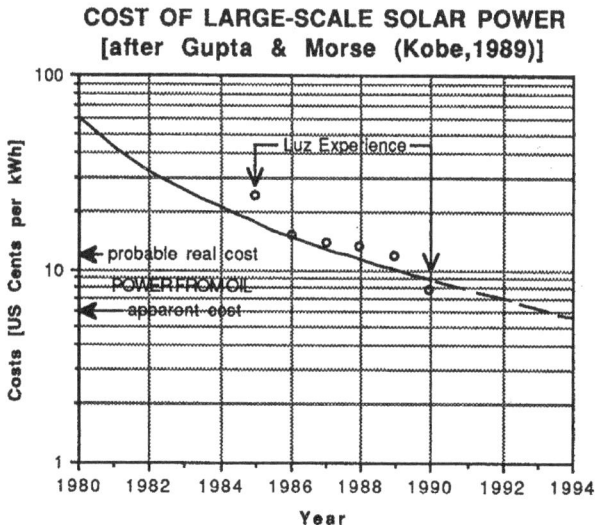

Figure 4.4.1: Decrease in reported SEGS cost vs. time

The graph is taken from the paper "Solar Thermal Technology Applications and Opportunities", by B.P. Gupta and F.H. Morse [published in *Clean and Safe Energy Forever* (Pergamon, New York, 1989) p. 1253]. Do not take the ordinate values too literally as this could initiate a side-tracking discussion on their accuracy instead of an appreciation of the real issues, namely that large-scale solar power (in sunny areas) from thermal plants is already competitive with energy from fossil-fuel-fired power stations if the *true* cost of this energy is calculated. Hohmeyer [2] and others have estimated that the social cost of conventional power adds 30–100% to the apparent cost of fossil power. The diagram also suggests very strongly (based on the two authors' extrapolated line) that, had the support for Luz continued for another two or three years, they might have reached, without requiring further subsidies, economic viability, when allowing only a minimal or even zero benefit for the energy being clean.

History may show this to have been a terrible tragedy, not just for the company and its employees but for the whole process of the acceptance of solar-energy as a viable alternative to fossil energy.

4.4.2.2 *Photovoltaic electricity as an alternative?*

As for PV electricity on a large scale, it is now admitted that the predictions were somewhat over-optimistic and this convenient conversion process does not yet compete with the thermal systems. However, at the intermediate level, costs are in an acceptable range and R&D to produce PV cells of higher efficiency and improved stability (which has been a problem with the cheaper amorphous Si cells) will reduce this still further. This means that the area of applicability of PV systems — and their economically viable size — will increase.

We can now consider 3 or 4 scenarios:

1) *Small Power-Supplies for Remote Areas in Developing Countries.*

   (a) It has been demonstrated that a small PV panel plus storage battery can provide evening lighting — to replace the proverbial oil lamp — and this encourages evening study — and a special lamp with high lumens per watt has been developed. This is an exact parallel to the early days of the petroleum industry — before the introduction of the internal combustion engine — where the first use (and market) was for lighting.

We are here faced with a policy problem that occurs over and over again, namely, that the payback on an investment in solar energy — in this case lighting — may take a generation and few finance ministers will support such a development. Yet the developed world, if it wants to live in peace, may have no alternative but to find a mechanism for supporting this type of development.

(b) The use of TV sets for educational purposes (as well as keeping people informed) is similar to the lighting application.

(c) Where solar-PV electricity can support some small-scale industrial activity — as an alternative to Diesel-power, the payback time may be much shorter — and here there is an area for entrepreneurial activity.

2) *For Larger Stand-alone Power-Supplies in Developing Areas,* which may not have a national grid but may — or could have — a local grid, solar-power, probably by thermodynamic conversion, is a serious option and its competitive status with local diesel power must be checked in each case and may prove positive even before the pollution costs (with their extended pay-back period) are added into the equations. For sunny areas, concentrating collectors — either of the Luz or central-receiver types for bigger (MW_e) units or dish-concentrators in the smaller (kW_e) ranges — are a possibility, though the technologies require further refinement. For less-sunny areas, where the performance of concentration-type collectors is poor, the solar-pond is a proven technology, provided that the site-conditions are appropriate. PV is not yet really ready for these stand-alone applications, on the grounds of cost.

3) For the Developed World. Luz has demonstrated — at the GW_e level — that large-scale solar-power in sunny areas is a viable option. If this energy does not at present appear to compete costwise with conventional energy, it is because the societal cost

of the latter is intentionally not included, nor its hidden subsidies. Where these costs are allowed for — which requires that the ecological lobby be stronger than the oil lobby — we can foresee very large solar thermal power stations in sunny areas, with the trapped solar-energy transported, by a variety of means, from desert areas to the areas of large consumption. The prognosis, for the next decade or two, for large-scale PV power is less encouraging: unlike the thermal systems, the unit costs do not decrease with size, and the apparent complexity of the thermal systems, somewhat frightening for small units, holds no fears for GW power-station engineers, whilst storage and back-up are today still easier for the thermal systems. However, PV is not ruled out in developed countries for consumers not reached by the national grid.

4.4.2.3 *Solar for displacing conventional generation*

This symposium has not dealt with the heating and cooling of buildings, but where this is achieved by active or passive solar, it is equivalent to the generation of large amounts of electricity because of the large amount of energy (as electricity or fuel) that goes into the heating and cooling of buildings in developed countries and, indeed, this application of solar-energy is considered by many as one of the major potential contributions to reduced dependence on fossil fuels and to reduced atmospheric pollution that solar can make on a world scale. For heating, the technology is reasonably well established (for new buildings, not for retrofit): for cooling, much work remains to be done to improve the technology, i.e. to reduce capital and O&M costs.

4.4.3 *Conclusions*

For Israel, which until the full development of its oil-shale program, has to import 97% of its energy sources (in the form of oil or coal), it is a pity that the Ministry of Energy and the Treasury chose not to see fit to encourage the erection of multi-MW commercial-type solar plants. This is not the forum to criticize the Ministry of

Energy, particularly as the Minister has honored us with his participation, but many citizens do not understand why, two proposals notwithstanding (one for a Luz-type and one for a solar-pond-type), solar power stations for Eilat have not materialized. (It should be noted that in each case the proposers undertook to find the necessary capital). Presumably, this was because of lack of agreement on the form or size of subsidy to be granted.

It has to be recognized that, at this stage, all renewable energy programs need some help, not so much in R&D budgets but in the decision to build operating plants. Thus the investment cost for wind farms in California dropped by a factor of five in 5 years (1981–86) simply by building many plants. A similar drop is reported from Denmark in its wind program. The US experience of Luz in thermal conversion, showed a similar clear lesson: by providing subsidies (in one form or another), the US authorities enabled the development of large and successful solar power-stations of ever decreasing cost. Solar-pond costs may also be expected to drop significantly as more units are built — and new developments are in the pipeline — but these plants must be built in Israel, as orders from abroad will not be forthcoming when it is seen that Israel itself does not use this technology. Thus large-scale use of solar electricity in Israel is not only a move in the right direction of reduced dependence on imported fuel (with its political implications) or of reduced pollution (and its long pay-back time) but a potential billion-dollar export business. We here note that the Ormat Company, by its diligent entrepreneurial pursuit of geothermal energy conversion, has demonstrated that it is possible to create a multi-million dollar energy export business.

It is my conviction that both common sense and business sense will prevail and that this country will become both a large-scale user and large-scale exporter of solar technology.

4.4.4 *References*

[1] Michael Lotker "Barriers to commercialization of large-scale solar electricity: Lessons learned from the Luz experience". Sandia Report SAND91-7014. Available on-line at: http://webapp1.dlib.

indiana.edu/virtual_disk_library/index.cgi/4298428/FID3353/m
92004851.pdf (visited 18.6.18).

[2] Olav Hohmeyer. "The social costs of electricity renewables versus
fossil and nuclear energy" *International Journal of Solar Energy*
11 (1992) 231–250.

4.5 The Present Costs of Solar Electricity (Prof. David Faiman)

A 10-minute paper by Professor David Faiman, Ben-Gurion University of the Negev, Sede Boqer Campus.

4.5.1 *Introduction*

Solar-generated electricity has been demonstrated on the megawatt scale by four distinct technologies: wind turbines; linear concentrators operating in the 300–400°C temperature range; solar ponds at temperatures below 100°C; and photovoltaic plants. I shall not discuss wind power since this is a qualitatively different subject in the sense that what the technology converts to electricity is not the solar energy itself but, rather, one of its by-products (the kinetic energy of air molecules). In the present paper I would like to make some comparisons between three kinds of technology that collect solar radiant flux and convert it to electric power.

A direct comparison, from what has been published in the literature, is rendered somewhat difficult on account of the fact that the various types of technology are to be found at different geographical locations and different financing schemes were involved in each case. For the purposes of the present discussion, therefore, I have adopted the following strategy. I shall first review the documented performance of the largest plants of each type. I shall then use these performance figures to compute how much annual energy each plant would give if they were all situated at a single (sunny) location. Lastly, I shall adopt a uniform and well-defined financing scheme in order to compare directly, in US¢/kWh, the relative economics of the three hypothetical plants.

4.5.1.1 *Photovoltaic plants*

The largest photovoltaic plants were built by Arcosolar Corp. in California. The oldest is the 1 MW plant at Hesperia which has provided data since 1983. This plant consists of 108 dual-axis trackers each of which carries approximately 95 m^2 of flat modules. Table 4.5.1 displays the performance figures of this plant for the (typical) year 1988 [1]. From Table 4.5.1 we arrive at the sobering conclusion that, whereas individual solar cells have been demonstrated with DC efficiencies in excess of 20% under laboratory conditions, the Hesperia power plant, under field conditions, exhibits a total AC efficiency of only 6.4%.

4.5.1.2 *Solar ponds*

The largest solar pond power plant that has thus far been operated is the 250,000 m^2 pond at Beit Ha'Aravah, Israel, designed, by Ormat Corp. to drive one of their 5 MW organic-fluid Rankine-cycle turbine generators. Although this pond was never allowed to drive the turbine at full power (owing to the fact that budgetary problems prevented inclusion of heat exchangers of the required size [2]), both the pond and turbine were operated for a sufficiently long period of time to enable assessment and publication [2, 3] of their performance. The observed [3] thermal efficiency of the pond was 16%, which compares quite favorably with the 20%, or so, that would be expected for a pond of the Beit Ha'Aravah dimensions if modeled according to

Table 4.5.1: Performance of the Arcosolar 1 MW photovoltaic plant at Hesperia, California, for the year 1988 [1].

Total aperture area [m^2]	10,275
Ground cover ratio	8.0
Total insolation [kWh/m^2]	3,241
Total dc output [MWh]	2,231
Mean dc efficiency [%]	6.7
dc $\rightarrow$ ac efficiency [%]	96

Ref. [4] under the simplistic assumptions that perfectly clean water and no bottom losses are assumed.

The power plant was observed [3] to have a thermodynamic efficiency of 5.5% (but 30% of the resulting electrical output had to be forfeited in the form of parasitic losses). The *planned* inlet and outlet temperatures across the turbine were respectively 28°C and 72°C from which one might expect a maximum efficiency of:

$$1 - \sqrt{(301/345)} = 0.066 \tag{4.1}$$

Since electricity is produced by an irreversible thermodynamic process, rather than by a Carnot-type engine — from which no power could be drawn in finite time, I have used the equivalent of Carnot's Law for finite-time thermodynamic processes [5] to make this estimate.

From Eq. (4.1) it is seen that the efficiency achieved by this 5 MW power plant is most reasonable for a first-of-a-kind example.

Combining the efficiencies of the pond and the turbine we see that the, total solar → electricity efficiency of the Beit Ha'Aravah solar pond plant is 0.88%.

4.5.1.3 *Linear concentrator plants*

By far the largest solar power plants that have yet been constructed are the 355 MW_e of Solar Electricity Generating Stations (SEGS) that Luz Corp built for Southern California Edison, in the neighborhood of Daggett, California. SEGSI was a 15MW prototype, SEGS2–SEGS7 are 30MW plants employing larger parabolic trough mirrors than the prototype and each of the latest 80 MW plants, SEGS8–SEGS9, has about 0.5 km^2 of still larger mirrors. The reflector troughs track the sun about a horizontal NS axis, heating oil to about 350°C at an annual average efficiency (in the documented [6] case of SEGS5) of 50%. This oil is then used to flash steam which drives a turbine at an efficiency of 30.6%. The solar-thermal efficiency of the Luz collectors compares favorably with measured efficiencies that have been published for a variety of parabolic trough collectors. For a power-producing engine operating at these temperatures one may expect a thermodynamic efficiency of about

$$1 - \sqrt{(292/623)} = 0.315 \tag{4.2}$$

but since the turbine employed is an off-the-shelf type the excellent value for the observed efficiency should come as no surprise.

We thus see that the total solar $\rightarrow$ electrical efficiency of the Luz linear concentrator plants is about 15%.

4.5.2 *Annual expected energy deliver at a single location*

In order to compare the relative performances of these three plants we must first "move them" to a single site. For this purpose I shall take Sede Boqer and I shall use (somewhat arbitrarily) insolation data recorded for the year 1988.

For an Arcosolar-type PV plant at Sede Boqer, the total global irradiance that would be received on a dual-axis tracking surface is 2,853 kWh/m^2. (This figure is derived from data on direct-beam and horizontal global insolation, using the isotropic assumption for the diffuse component and an assumed desert albedo of 0.3). Under these conditions and at an assumed efficiency of 6.4% this 10,275 m^2 photovoltaic system would produce 1.88 GWh of AC electricity per year. Taking into consideration the required ground cover ratio of 8:1 for the dual-axis trackers, each square kilometer of land could generate 22.8 GWh of electrical energy.

For an Ormat-type solar pond plant located at Sede Boqer, the annual global horizontal irradiance is 2087 kWh/m^2. This figure should be reduced by an annual average incidence angle modifier of about 0.95 (to allow for the fact that not all of the incoming radiation enters perpendicularly to the pond surface). Thus at a total solar $\rightarrow$ electricity efficiency of 0.62% (this includes the 30% parasitic loss mentioned above) the 250,000 m^2 Beit Ha'Aravah solar pond, if moved to Sede Boqer, would produce 3.07 GWh of AC electricity per year. Since the ground cover ratio is essentially 1:1 for large ponds, each square kilometer of land could generate 12.3 GWh of electrical energy.

In the case of a Luz-type plant at Sede Boqer the calculation is, of needs, more artificial. In real life these plants operate in conjunction with a gas-fired backup system that ensures the rated plant output during periods of the day when electricity revenues are highest. The actual strategy for use of the backup gas is quite complicated since local regulations limit the amount of gas that may be consumed by a hybrid solar/fossil-backed plant. In fact, such plants cannot be operated without backup unless one is prepared to suffer a situation of no electrical output all the time the oil is heating up in the morning (about an hour [7]) and, more generally, whenever the radiation drops below the collector's threshold. For the present calculations I have artificially assumed that backup gas is used, when necessary, merely to maintain the oil temperature at its threshold value from sunrise till sunset. In this way the collectors are able to convert all of the incoming direct beam radiation to electricity. For Sede Boqer the total direct beam irradiance on a horizontal tracker with NS-axis would be 2001 kWh/m^2 per year. For this kind of tracking strategy the annual correction due to incidence angle modifier is small (2–3%) and I have chosen to neglect it. Thus at an efficiency of 15% the 233,120 m^2 of SEGS5 collectors would produce 70.0 GWh of solar electricity. Scaling this figure to 1 km^2 of land and allowing a 2.5:1 ground cover ratio [6] for these linear concentrator collectors we arrive at a total annual generating capacity of 120 GWh per square kilometer of land.

4.5.3 *Economic comparisons*

Although the cost of the Hesperia system was never revealed I shall assume it to have been US$10 per peak watt — typical of system costs at that time. (This would have corresponded to US$5 per watt for the photovoltaic modules and US$5 per watt for the balance of system (BOS) costs). Thus the 1MW Hesperia system cost is taken here as US$10M. I shall also assume a uniform financing scheme for all three plants, viz, a 20 year loan at 8% p.a. interest: This corresponds to Rabl's "10% rule" [8], i.e. the 1.8 GWh of annual electricity, that such a plant would produce at Sede Boqer, would cost US$1M to produce. This amounts to 53US¢/kWh to which one may add the reported [9] annual operation and maintenance costs of 0.3 US¢/kWh.

The capital cost for the 250,000 m^2 solar pond power station at Beit Ha'Aravah was [3] US$20.18/m^2. Using the same financing scheme we assumed for the photovoltaic system the 3.07 GWh of annual electrical output that an Ormat-type pond plant would provide at Sede Boqer would cost 16.4 US¢/kWh. To this figure should be added annual O&M costs of US$510,000 [3], i.e. another 16.6 US¢/kWh, making a total of 33 US¢/kWh.

Lastly, the price of the Luz SEGS5 plant was about US$110M [6]. This implies a cost of 15.7 US¢/kWh for the 70 GWh of annual electrical output that such a plant would yield at Sede Boqer. O&M costs amount to about 2% of the system cost per year [6]. This adds another 3.1 US¢/kWh to the cost of electricity, making a total of 18.8 US¢/kWh. But in point of fact this is an underestimate as it assumes zero cost for the backup fuel.

4.5.4 *Conclusions and outlook*

From the above considerations we arrived at the result that present day solar power generating technology can generate electricity at a cost of about 53 US¢/kWh in the case of photovoltaics, at about 33 US¢/kWh in the case of solar ponds, and at about 19 US¢/kWh in the case of parabolic trough systems.

In the photovoltaics example we have chosen, half of the 53 US¢/kWh represents module costs and half are BOS costs. It has been estimated [10] that BOS costs might realistically be reduced by a factor of 5 compared to the assumed US$5 per peak watt we took for our Sede Boqer example. Such reduced BOS costs would only contribute 5.3 US¢/kWh to the cost of the electricity produced by the system. If the total electricity cost is to be no larger than, say, 10 US¢/kWh then the module cost must be reduced by a factor of between 5 and 6. Alternatively, module costs per m^2 could remain at the level assumed here if the system AC efficiency would increase to 36%.

Without wishing to enter here into the controversial matter of the "true cost" of environmentally acceptable electric power, one sees that considerable technological improvement will be needed if photovoltaics are to compete, on a large scale, with conventional power stations.

In the case of solar ponds, Ref [3] points out that the same 5MW turbine that Ormat built for the 250,000 m^2 pond at Beit Ha'Aravah could simultaneously serve four such ponds. In such a case the 30% parasitic loss mentioned above could probably be reduced to about 10% [11], and the thermodynamic efficiency of power generation could rise to the 6.6% figure computed above for the pond design temperatures. It is also reasonable to suppose that the pond thermal efficiency could be made to rise from its prototype 16% to the design value of 18%. Under such circumstances a 1km^2 system would produce 21.2 GWh per year (at Sede Boqer). According to ref [3] the system cost would then fall to US$10.37/m^2, contributing 4.9 US¢/kWh to the cost of delivered electrical energy. The annual O&M costs of [3] US$900,000 would contribute a further 4.2 US¢/kWh making a total of 9.1 US¢/kWh for the electricity.

It appears, therefore, that in the case of solar ponds, its continued development rather than a technological breakthrough, is needed to produce economically competitive solar power on a large scale.

In the case of high temperature solar-thermal systems it is not clear whether the cost of the solar part of the electricity can be brought down much below the 18 US¢/kWh level. On the other hand, any kind of plant which could literally be switched off by each passing cloud (owing to its dependence on direct beam radiation) would obviously not be operated as a purely solar plant. The hypothetical calculations presented here assumed cost-free gas burned in a highly arbitrary manner — namely, so as to keep the plant temperature above threshold, all day and all year. Luz, of course, uses a more practical strategy (one which guarantees full plant output during peak hours, whatever the weather) and its claim is that the electricity from its SEGS-8 and SEGS-9 plants works out at under 10 US¢/kWh. The extent to which the cost of electricity from hybrid linear concentrating systems (such as the Luz plants), or from any of the potentially higher-efficiency concentrating configurations (that have not yet been demonstrated in the commercial sector), can be made directly competitively with conventionally generated electric power is not clear. On the one hand, point-focusing systems (from which higher efficiencies may in principle be obtained) have not yet been successfully demonstrated in the commercial sector. So the potential cost of electricity from such plants is still very much conjectural. On the other hand, the price of electricity from linear concentrators has fallen steadily with each new generation of Luz plants. So perhaps all that is needed for solar thermal electricity (both troughs and ponds) to compete on the open market with fossil fuel is continued growth.

4.5.5 *References*

[1] V.V. Risser, personal communication, Feb 1, 1991.

[2] B. Doron, Summary of one year of operation of the 5MW project at Bet Ha'Aravah, in Proc. of the 1st Sede Boqer Workshop on Solar Electricity Production. 23–24 February 1986, ed. D. Faiman (Jacob Blaustein Inst for Desert Research Pub # ASCU-86/15) pp H23–H25 [in Hebrew. Authorized English translation available from the editor]. [*Editorial comment: The translation referred to here is reproduced in chapter 1 of the present book.*]

[3] H.Z. Tabor and B. Doron, "The Beith Ha'Arava 5MW(e) solar pond power plant (SPPP) — Progress Report". *Solar Energy* **45** (1990) 247.

[4] C.F. Kooi, "The steady state salt gradient solar pond". *Solar Energy* **23** (1979) 37.

[5] F.L. Curzon and B. Ahlborn, "Efficiency of a Carnot engine at maximum power output", *American J. Physics* **43** (1975) 22.

[6] J.F. Kreider, Performance of a large solar-thermal electric power plant in the USA, in Proc. of the 3rd Sede Boqer Symposium on Solar Electricity Production. 6–7 March 1988, ed. D. Faiman (Jacob Blaustein Inst for Desert Research Pub # ASCU-88/19) pp 53–85. [*Editorial comment: This reference is reproduced in chapter 3 of the present book.*]

[7] Luz, personal communication Feb 6, 1991.

[8] A. Rabl, *Active Solar Collectors and Their Applications*, (Oxford University Press, New York, etc, 1985)

[9] A.L. Rosenthal, Photovoltaic System Performance Assessment for 1989. EPRI report # GS-7286 (Interim report May 1991) pp 6–14.

[10] E. Berman, Large photovoltaic power plants, in Proc. of the 1st Sede Boqer Workshop on Solar Electricity Production. 23–24 February 1986, ed. D. Faiman (Jacob Blaustein Inst for Desert Research Pub # ASCU-86/15) pp E79–E80.

[11] Ormat, personal communication, Feb 17, 1991.

Chapter 5

1993

5.1 Editor's Foreword

A solar tower, or "central receiver", power plant consists of a power-producing structure — typically some kind of boiler — mounted atop a tower, to which the Sun's rays are directed by a field of Sun-tracking mirrors, — so-called "heliostats". During the late 1970's and early 1980's a number of such systems were constructed and tested in various parts of the world. Their performance was monitored for a few years and documented in a special issue of the *ASME Journal of Solar Energy Engineering* that was dedicated to this kind of technology. These systems were in France [1], Italy [2], Japan [3], two in Spain [4, 5], and the USA [6]. The largest was the 10 MW *Solar-One* system at Barstow in California. At a later stage, based upon knowledge gained at *Solar-One*, this system was converted into *Solar Two*. The keynote presentation of Paul Klimas was a state-of-the-art review of solar tower knowledge at the time of that conversion.

The keynote review of the development of photovoltaic cells in the Soviet Union, delivered at this symposium by Mark Koltun, will appear in volume 2.

5.1.1 *References*

[1] L.P. Drouot and M.J. Hillairet, "The Themis Program and the 2500-KW Themis Solar Power Station at Targasonne" *ASME Journal of Solar Energy Engineering* **6** (1984) 83–89.

[2] D. Borgese, G. Dinelli, J.J. Faure, J. Gretz and G. Schober, "Eurelios, The 1-MW(el) Helioelectric Power Plant of the

European Community Program" *ASME Journal of Solar Energy Engineering* **6** (1984) 66–77.

[3] T. Hirono and T. Horigome, "A 1-MWe Central Receiver Type Solar Thermal Electric Pilot Plant" *ASME Journal of Solar Energy Engineering* **6** (1984) 90–97.

[4] W. Grasse and M. Becker, "Central Receiver System (CRS) in the Small Solar Power Systems Project (SSPS) of the International Energy Agency (IEA)" *ASME Journal of Solar Energy Engineering* **6** (1984) 59–65.

[5] A. Munoz Torralbo, C. Hernandez Gonzalvez, C. Ortiz Roses, J. Avellaner Lacal and F. Sanchez, "A Spanish 'Power Tower' Solar System: Project CESA-1" *ASME Journal of Solar Energy Engineering* **6** (1984) 78–82.

[6] J.J. Bartel and P.E. Skvarna, "10-MWe Solar Thermal Central Receiver Pilot Plant" *ASME Journal of Solar Energy Engineering* **6** (1984) 50–58.

5.2 The Solar-Two Power Tower Project (Dr. Paul Klimas)

A keynote lecture presented by Paul C. Klimas* (Sandia National Laboratories, Albuquerque, NM, USA). Please note that although this invited keynote lecture was presented by Dr. Klimas, the paper that he provided for publication in the Symposium Proceedings — which is reproduced here — was co-authored by: James M. Chavez (Sandia Labs), Pascal De Laquil III (Bechtel Corporation, San Francisco, CA), and Mark Skowronski (Southern California Edison, Los Angeles, CA).

5.2.1 *Abstract*

A consortium of United States utility concerns led by Southern California Edison Company (SCE) has begun a cooperative project with the U.S. Department of Energy (DOE) and industry to convert the 10 MWe *Solar-One* Power Tower Pilot Plant to molten nitrate salt technology. Successful operation of the converted plant, to be called *Solar-Two*, will reduce the economic risks in building the initial commercial power tower projects and accelerate the commercial acceptance of this promising renewable energy technology. In a molten salt power tower power plant, sunlight is concentrated by a field of sun-tracking mirrors, called heliostats, onto a centrally located receiver, atop a tower. Molten salt is heated in the receiver and stored until it is needed to generate steam to power a conventional turbine generator.

Joining SCE and DOE in sponsoring this project are the following organizations: Los Angeles Department of Water and Power, Idaho Power Company, PacifiCorp, Pacific Gas and Electric Company, Sacramento Municipal Utility District, Arizona Public Service Company, Salt River Project, City of Pasadena, California Energy Commission, Electric Power Research Institute, South Coast Air Quality Management District, and Bechtel Corporation. The *Solar-Two* project will convert the *Solar-One* heat transfer system from a water/steam type to molten nitrate salt by replacing the water/steam receiver and oil/rock thermal storage systems with a nitrate salt receiver, salt thermal storage, and steam generator. The estimated cost of *Solar-Two*, including its 3-year test period, is US\$ 48.5 million. The plant will be on line in early 1995.

Keywords: Solar thermal energy, *Solar-Two*, Power towers, Receivers, Solar power plants, Nitrate salts, *Solar-One*.

5.2.2 *Introduction*

A solar power tower plant uses a field of sun-tracking mirrors, called heliostats, to concentrate sunlight onto a receiver, where the thermal energy is collected in the form of a heated fluid. The concept, shown in Fig. 5.2.1, was first proposed by scientists in the USSR in the mid-1950s. In the past 18 years, a number of component experiments and system experiments have been fielded around the world, to demonstrate the engineering feasibility of the concept and validate its potential.

The 10 MWe *Solar-One* Pilot plant near Barstow, California, shown in Fig. 5.2.2, was the largest of the system experiments. *Solar-One* established the technical feasibility of the power tower concept. Initiated in 1976, the plant used a water/steam receiver coupled directly to the turbine-generator. A thermocline oil/rock system provided thermal storage. This approach was selected to

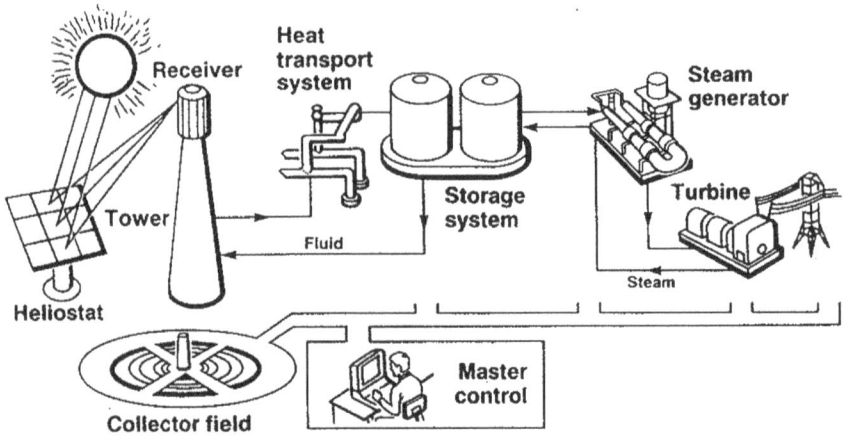

Figure 5.2.1:   Solar power tower concept

Figure 5.2.2:   *Solar-One* pilot plant [PIX No. 00036, Sandia Natl. Labs., NREL, U.S. Dept. of Energy]

establish utility confidence in the concept and to minimize technical risks. *Solar-One* used 1,818 glass metal heliostats to reflect the solar energy to the receiver on the tower. *Solar-One* started operation in 1982, and its 6-year test and power production program proved that the technology operates reliably, has very low environmental impacts, and excellent public acceptance [1]. The two key disadvantages of the water/steam system at *Solar-One* were that the receiver was directly coupled with the turbine, causing the turbine to trip each time a cloud came by, and use of the oil/rock thermal storage system was not efficient because of thermodynamic losses [2]. After successfully completing its test and operation, *Solar-One* was decommissioned in 1988 [3].

In parallel with *Solar-One*, a series of studies, funded by the U.S. Department of Energy and Industry, examined advanced central receiver concepts that used a single-phase receiver fluid [4, 5]. Both molten nitrate salts and sodium were considered. The advantages of the single-phase fluid for a solar power tower plant include:

- The receiver can be operated at a lower pressure and higher incident flux than a water/steam receiver. This translates into a smaller, more efficient, and lower cost receiver and support tower.
- The single-phase fluid can be stored in a large tank at atmospheric pressure. This allows the plant to 1) isolate the turbine-generator from solar energy transients, 2) economically and efficiently store thermal energy collected early in the day for use during peak demand periods, 3) increase the plant capacity factor by oversizing the collector and receiver systems and storing the excess thermal energy for electricity generation in the evening, and 4) operate the turbine at maximum efficiency.

The interest in sodium as a heat transfer fluid waned because of its inherent safety problems. However, development of power tower systems using molten salt continued because of salt's advantages. Although the power tower plants using a singlephase fluid offer significant technical and economic improvements over the water/steam plants [6], earlier attempts [7] to retrofit *Solar-One* with a molten nitrate salt heat transfer system failed. Even though federal and state tax credits were available, the large technical and financial risks to the investors, decreasing fossil fuel prices, and excess utility capacity precluded project financing. Also, there was no immediate or intermediate market seen for power tower plants. Nevertheless, confidence in the future of advanced power tower plants continued.

Efforts by DOE, Sandia National Laboratories, and industry to demonstrate nitrate salt components and a system experiment have been technically successful. The component tests included two $5\,MW_{th}$ receivers, a $7\,MWh_{th}$ thermal storage system, a $3\,MW_{th}$ steam generator, and a pump and valve loop sized for a $190\,MW_{th}$ receiver [8]. The Molten Salt System Experiment used one of the receivers, the thermal storage tanks, and steam generator from the component tests and a $750\,kW_e$ turbine-generator to form a complete nitrate salt system. However, the size of the system experiment was not large enough to demonstrate the economic potential of a commercial facility [9]. A schematic of the molten salt central receiver power plant technology is shown in Fig. 5.2.3.

The next step in the commercialization of central receiver technology is to design, construct, and operate a demonstration plant of a size that is large enough to reduce to acceptable levels the risks in building the first commercial plant. *Solar-Two* is this plant.

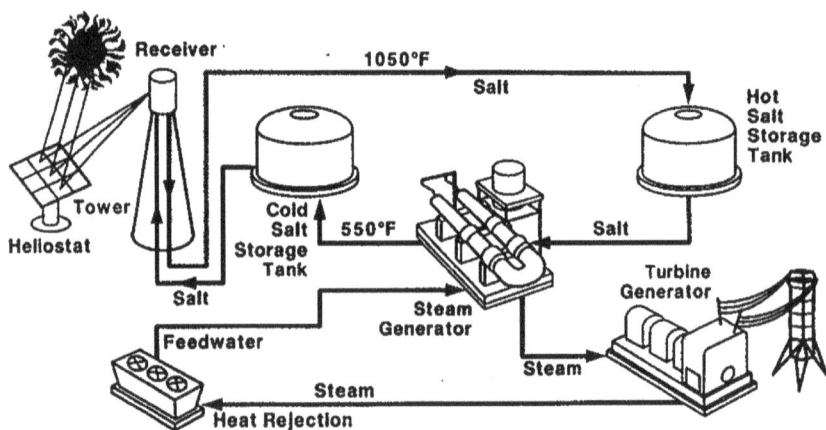

Figure 5.2.3:   Schematic of molten salt system flow

5.2.3 *Solar-Two: Scope, objectives and risk mitigation*

The *Solar-Two* Project will require the removal of *Solar-One*'s entire water/steam heat transfer system, including the receiver, piping, and heat exchangers, and the oil/rock thermal storage system. The *Solar-One* power block, including the heat transfer system and thermal storage system are shown from above in Fig. 5.2.4. The new molten salt system to be installed will include a receiver, a dual tank thermal storage system, and a steam generator system. Additional heliostats will be added to the south part of the field, and the turbine will be upgraded to improve the plant performance. Other

Figure 5.2.4:   *Solar-One* power block (zoom-in from Fig. 5.2.2)

minor improvements will be made, and damaged equipment will be replaced. However, the balance of the plant will be utilized in *Solar-Two*. A detailed description of the modifications present in *Solar-Two* is provided in Section 5.2.4.

The principal goal of the *Solar-Two* project is to significantly reduce the perceived technical and economic risks in building the first commercial projects. To accomplish its goal, the project has established the following objectives:

- Validate the technical characteristics of the nitrate salt receiver and storage technology.
- Improve the accuracy of economic projections for commercial projects by increasing the data base of capital, operating, and maintenance costs.
- Simulate the design, construction, and operation of the first commercial plants.
- Collect, evaluate, and distribute to United States utilities and the solar industry the knowledge gained from the project to foster wider utility interest in commercial plants.
- Stimulate the formation of a commercialization consortium that will facilitate the construction of the initial commercial projects.

The project will address the perceived technical and economic risks in the following manner:

- Plant Availability. During its 3-year power production phase, the *Solar-One* pilot plant had annual availabilities of 80 to 83 percent, and during its last year under SCE operation, an availability of 96 percent [1]. Commercial central receiver plants require an availability of 90 percent to achieve published energy cost goals. The 3-year operation of *Solar-Two* will provide the required data for high confidence in the commercial plant projections.
- Plant Design Point Performance. The largest uncertainties in design point performance are the receiver spillage and convection losses; the optical and thermal performances in the balance of the plant are well understood. Data from an improved beam

characterization system and receiver thermal loss measurements during the 1-year test period will be available to corroborate or refine these loss models.

- **Annual Performance.** Annual plant performance is estimated with the SOLERGY computer code, which uses equipment performance characteristics and site weather data to calculate energy flows at either 3- or 15-minute intervals during the year. Although the accuracy of the code has been corroborated by comparing the annual performance of the *Solar-One* plant with computer projections [10], the 2-year power production phase of the *Solar-Two* project will provide an opportunity to verify the accuracy of the code for nitrate salt systems and to increase confidence in projections for commercial projects.

- **Annual Operation and Maintenance (O&M) Cost.** O&M expenses for the *Solar-One* plant are well understood, and projections of O&M expenses for the *Solar-Two* and commercial projects are believed to be accurate. However, operation of the *Solar-Two* project will provide an opportunity to validate the projections for the $10\,\text{MW}_e$ plant and to increase confidence in the estimates for the commercial projects.

- **Equipment Prices and Warranties.** This is one of the largest risks. Suppliers of the equipment for the initial commercial projects will be required to provide competitive bids and performance warranties. The *Solar-Two* project will provide an opportunity for vendors to design, fabricate, and test equipment at a size within a factor of 3 of that required for the first commercial project. This experience, and a summary of the non-proprietary information and lessons learned, will be distributed to the central receiver community and will reduce the uncertainties in providing equipment for the first commercial projects. The reduced risks, together with competitive bids, will help to minimize equipment prices on the initial projects.

- **System Performance Guarantees.** To finance construction of the initial $100\,\text{MW}_e$ commercial projects, a limited annual performance guarantee will likely be required to minimize risk to the investors. Successful operation of the *Solar-Two* project will reduce the system performance uncertainties and allow a guarantee to be offered for the initial commercial projects.

5.2.4   *Description of solar-two*

The basic plant configuration, project organization, schedule, cost estimate, and funding sources for the *Solar-Two* project have all been identified.

5.2.4.1   *Plant configuration*

As stated previously, the *Solar-Two* project replaces the previous steam-based receiver, heat-exchangers, piping and oil/rock storage, with a new heat transfer system based on nitrate salt. Other than that, it utilizes the existing, albeit slightly enlarged, heliostat field, turbine-generator plant, receiver tower, and a portion of the master control system from the *Solar-One* plant. A preliminary piping and instrumentation drawing for *Solar-Two* is shown in Fig. 5.2.5. One of the project objectives is to simulate the 100 MW plant designs. To meet that objective, the project will have the following features:

- Approximately 11,500 m^2 of additional mirror area will be added to the south side of the collector field to provide a flux distribution representative of a commercial receiver, eliminate excessive morning start times common to the *Solar-One* plant, and provide additional energy for charging the thermal storage system. Of the additional mirror area, current plans call for approximately 10,000 m^2 to be flat glass on refurbished two-axis trackers and 1,500 m^2 will be new stretched-membrane heliostats for research purposes.
- A 43 MW$_{th}$ external cylindrical nitrate salt receiver will replace the present water/steam receiver. The salt receiver will be 6.1 m in diameter and 6.6 m tall, and will use grade316 stainless steel tubes with 1.9 cm, od and 0.17 cm wall. The molten salt will enter the receiver at 285°C, flow in two control zones in a serpentine path through the receiver, and exit the receiver at 565°C. The average incident flux on the receiver will be 0.30 MW$_{th}$/m^2.

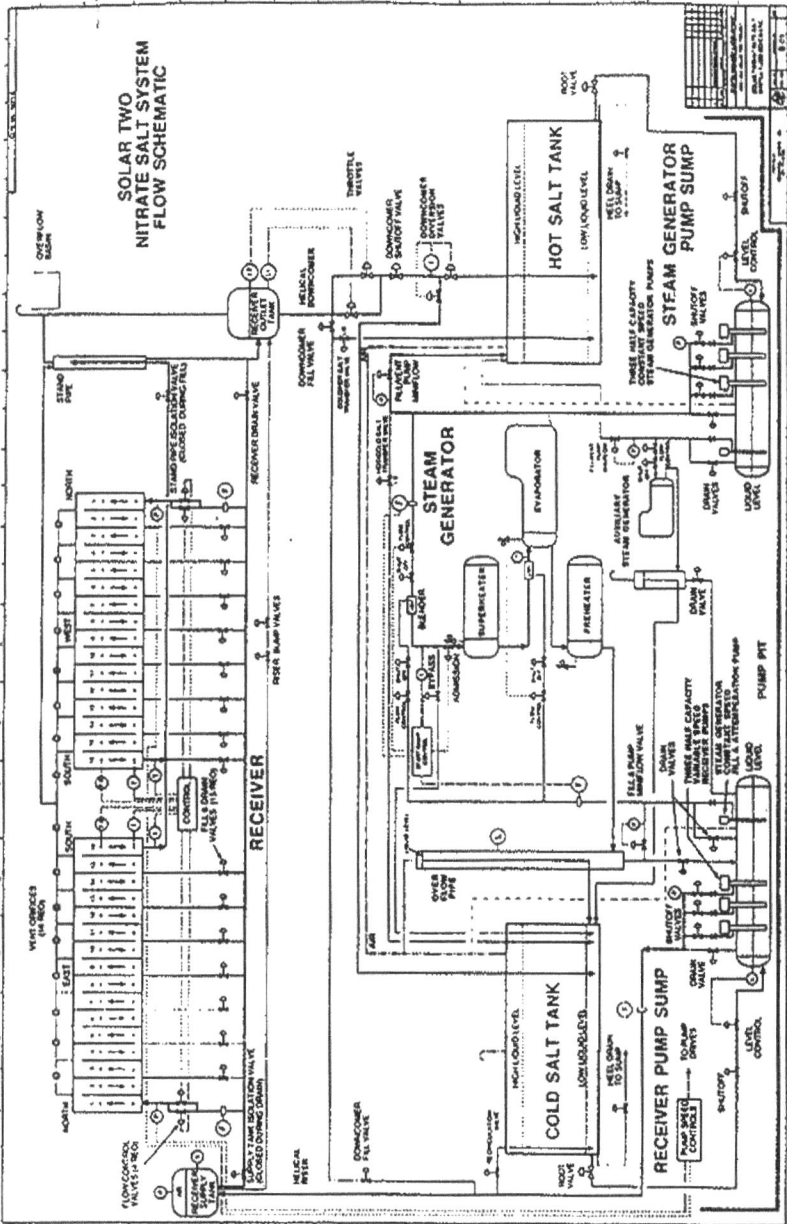

Figure 5.2.5: *Solar-Two piping and instrumentation drawing*

- A nitrate salt thermal storage system will replace the present oil/rock thermocline unit. The system will use separate hot (565°C) and cold (285°C) salt tanks and will be sized to operate the turbine at full load for 4 hours. Both tanks will be on the order of 13.7-m diameter and 6.1-m tall. Over a million kilograms of salt will be used in this system. The pumps for pumping the cold salt up to the receiver will be multistage vertical turbine pumps. The hot pumps, to pump the salt to the steam generator, will be vertical cantilever pumps. The storage system will demonstrate the decoupling of solar energy collection from the generation of electrical energy, the potential to meet a utility's evening peak demand, and the dispatch characteristics of a commercial plant.
- A 35 MW$_{th}$ steam generator will be added to convert the thermal energy in the nitrate salt to 512°C steam for the turbine-generator. The steam generator will consist of u-tube/u-shell superheaters (two shells in parallel) and preheaters. The evaporator will either be a kettle boiler or a u-tube/u-shell design.
- Modifications to the turbine plant will be made to improve its thermal efficiency and reduce its auxiliary electrical energy use. The addition of a fifth feedwater heater will increase the turbine cycle efficiency by 1 percentage point to 36 percent.
- Modifications and additions to the master control system will be made to replace equipment no longer supported by the original suppliers and to integrate the existing and new heliostats under one heliostat array controller.

5.2.4.2 *Schedule*

A preliminary project schedule, shown in Fig. 5.2.6, includes a 1-year period for final design, 1 year for equipment fabrication and installation, 6 months for startup and checkout, 1 year for system test and evaluation, and 2 years of power production to simulate a commercial plant. The project is planned to begin its test period in the second quarter of 1995 and its power production phase in the second quarter of 1996.

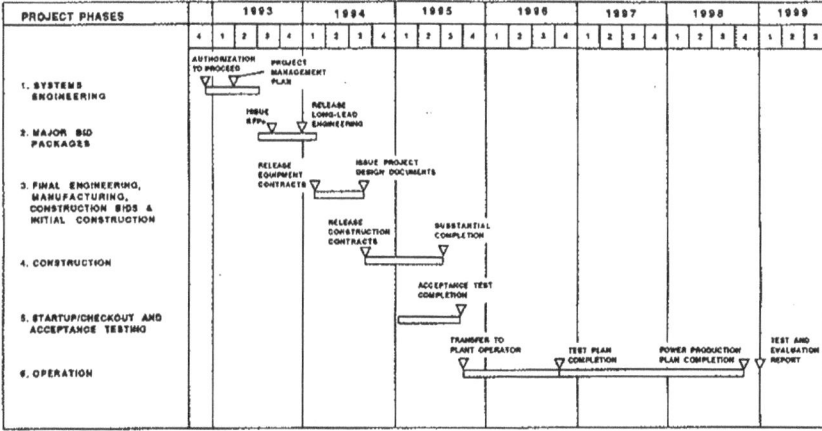

PROJECT PHASES	1993	1994	1995	1996	1997	1998	1999

Figure 5.2.6:   *Solar-Two* design and construction schedule

5.2.4.3   *Funding*

A summary of the total project cost estimate is shown in Table 5.2.1. The overall cost of US$ 48.5 million includes all expenses for final design, construction, startup and checkout, and annual operation and maintenance during the 1-year test period and the 2-year power production phase.

The principal project participants and their financial contributions are presented in Table 5.2.2. The capital and operation and maintenance costs are to be shared on a 50–50 basis between DOE and the consortium sponsors. This equal cost-sharing between the Federal government and its partners is typical of the Clinton administration's approach to furthering US technologies.

5.2.4.4   *Solar-two project organization*

The *Solar-Two* project organization is shown in Fig. 5.2.7. All the utility concerns joined together to form a consortium under a participant's agreement. SCE, as the consortium lead, prepared a request for cooperation with the U.S. DOE under a Cooperative Agreement. Within the utility consortium, a contribution of

Table 5.2.1:   Project cost estimate

Item	Cost estimate [US$ K]
Structures and Improvements	1,100
Collector System	2,900
Receiver System	11,000
Thermal Storage System	4,800
Steam Generation System	2,500
Electric Power Generation System	3,100
Master Control System	2,200
Total Field Cost	**27,600**
Engineering and Construction Management	6,700
Project Management	2,300
Startup, Checkout, and Spare Parts	2,000
Overheads	400
Total Capital Cost	**39,000**
Operation and Maintenance	
Test and Evaluation (one year)	3,500
Power Production Phase (First Year)	3,000
Power Production Phase (Second Year)	3,000
Total Project Cost	**48,500**

Table 5.2.2:   Project funding commitments

Organization	Commitment [US$ K]
Southern California Edison	11,148
Los Angeles Department of Water and Power	1,261
Idaho Power Company	1,261
PacifiCorp	1,261
Pacific Gas and Electric Company	1,261
Sacramento Municipal Utility District	1,261
Arizona Public Service Company	630
Salt River Project	631
City of Pasadena	100
California Energy Commission	1,000
Electric Power Research Institute	200
Bechtel Corporation	1,750
Industry	2,071
South Coast Air Quality Management District	100
Sales from Net Power	315
Subtotal	**24,250**
U.S. Department of Energy	24,250
Total Project Cost	**48,500**

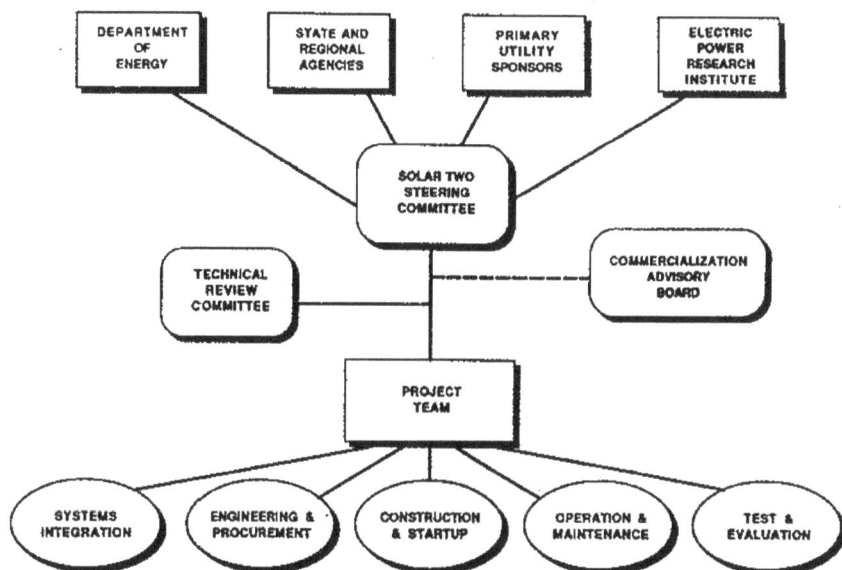

Figure 5.2.7: *Solar-Two* project organization

$1 million or more entitles an organization to place one representative on a steering committee. This committee sets project policy and objectives, and assures the accomplishment of the objectives as permitted by the project resources. Additional groups within the project organization include the Technical Advisory Committee, Commercialization Advisory Board, and the Project Team. The Technical Advisory Committee will review all designs and proposals and transfer information to all participants. The Commercialization Advisory Board will work to use *Solar-Two* as the stepping stone for the commercialization of power tower plants.

5.2.5 Commercialization plan

A plan for the commercialization of the technology, illustrated in Fig. 5.2.8, encompasses the following phases:

- Phase 1: Formation of an R&D Consortium for the *Solar-Two* project.

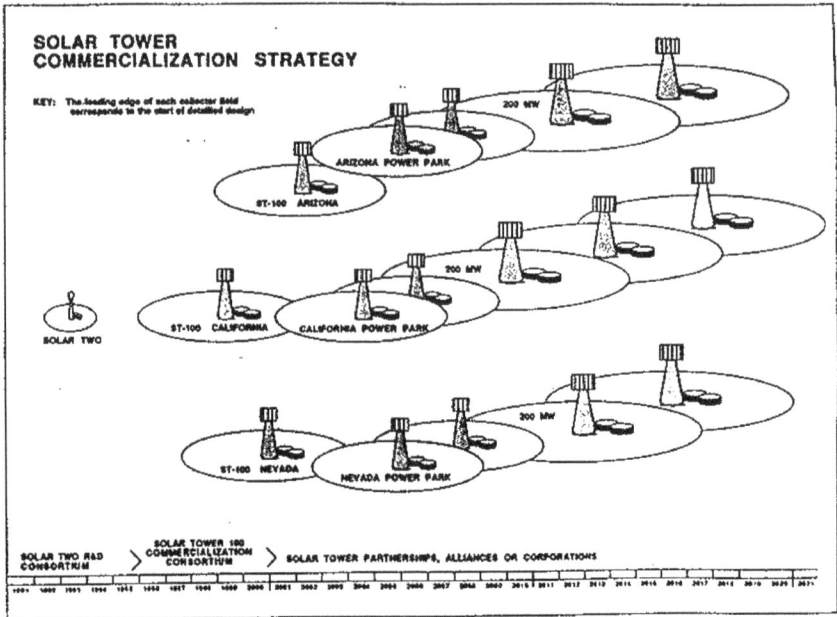

Figure 5.2.8:   Solar Tower commercialization strategy

- Phase 2: Expansion of the R&D Consortium into a Commercialization Consortium, which will facilitate the construction of the initial 100 MW$_e$ projects.
- Phase 3: Formation, from the Commercialization Consortium, of one or more commercial alliances, partnerships, or corporations that will market follow-on 100 MW$_e$ and 200 MW$_e$ plants in the United States and internationally.

The R&D Consortium for the *Solar-Two* project has been formed and is proceeding with design and construction.

During Phase 2, a Commercialization Consortium will be formed from an industry joint venture establishing the terms for constructing the initial plants, utility companies providing commitments for the initial plant orders, and investment groups providing debt and equity financing. The "initial" plants are defined as the first three to five 100 MW$_e$ projects, and their design will be based largely on experience from the 10 MW_e *Solar-Two* project. The key to reducing the plant capital cost to levels that will be acceptable to investors is to establish a steady demand for the most expensive items in the plants; specifically, the heliostats. One potential approach is to develop commitment agreements among one or more investors and interested utilities to build a minimum of three 100 MW_e projects as the initial plants. The plants would be built in succession at one or more utility sites at approximately 1-year intervals, and would establish an initial heliostat production rate of 3,000 per year. This should reduce heliostat prices from approximately US$ 180 m^{-2} for a single 100 MW_e project to US$ 120 m^{-2}, and thereby reduce the cost of energy to a value of US$ 0.11 per kWh_e for these initial plants.

As currently envisaged, the initial plants will be financed and operated on a commercial basis and will not require direct subsidies or financial guarantees from DOE or other public agencies. However, this does not rule out the potential for taking advantage of Federal and state programs for promoting new renewable energy technologies, such as investment or production tax credits and demonstration project rate structures. In particular, the ER-90 legislation in California provides a mechanism for promoting the first commercial-scale power plants (>50 MW_e) using advanced, clean technologies that could be utilized in developing and financing one of the initial plants.

During Phase 3, the rights, expertise, and knowhow developed by the industry joint venture through construction of the initial plants would be converted into one or more alliances, partnerships, or other ventures to develop, construct, and operate follow-on 100 MW_e and 200 MW_e projects. Utility and investor participants in the Commercialization Consortium could be part owners and investors in the new entity or entities, which would compete within whatever market structures exist for the construction of new generating capacity by United States utilities and internationally. It is anticipated that levelized energy costs of electricity from these higher volume production facilities would drop to approximately US$ 0.08 per kWh_e.

5.2.6   Conclusions

The *Solar-Two* project is well defined and timely. It is the first step in the commercialization process recommended at the conclusion of the utility studies [4, 5], it enjoys strong financial support from the USDOE and utilities in the western United States, and is well timed to meet the needs for large-scale renewable energy generation early in the next century.

5.2.7   *References*

[1] L.G. Radosevich, *Final Report on the Power Production Phase of the 10 MWe Solar Thermal Central Receiver Pilot Plant,* Sandia National Laboratories, SAND 87-8022, March 1988.

[2] S.E. Faas, *et al., 10 MW_e Solar Thermal Central Receiver Pilot Plant: Thermal Storage Subsystem Evaluation — Final* Report, Sandia National Laboratories, SAND 86-8212, 1986.

[3] G.J. Kolb and C.W. Lopez, *Reliability of the* Solar-One *Plant During the Power Production Phase,* Sandia National Laboratories, SAND 88-2664, 1988.

[4] T. Hillesland, Jr., (Pacific Gas and Electric Company, San Ramon, California), *Solar Central Receiver Technology Advancement for Electric Utility Applications — Phase I Topical Report,* PG&E Contract GM 633022-9, DOE Contract DE-FC04-86AL38740, and EPRI Contract RP 1478-1, August 1988.

[5] Arizona Public Service Company, *Arizona Public Service Utility Solar Central Receiver Study,* APS, Phoenix, 1988.

[6] P.K. Falcone, *A Handbook for Solar Central Receiver Design,* Sandia National Laboratories, SAND 86-8009, 1986.

[7] Pacific Gas & Electric, *Solar Central Receiver Technology Advancement for Electric Utility Applications — Phase II C Topical Report,* Report 007.25-92.2, 1989.

[8] Sandia National Laboratories, *Today's Solar Power Towers,* Sandia National Laboratories, SAND 91-2018, 1991.

[9] R.J. Roll, *Molten Salt Solar-Electric Experiment, Vol. 1: Testing, Operation, and Evaluation,* GS-6577, Vol. I, Research Project 2302-2, Electric Power Research Institute, 1989.

5.3   Discussion Following Paul Klimas's Presentation

Yogev: What is the total mirror area for the 10 MW$_e$ plant?

Klimas: The original *Solar-One* heliostat field comprised about 72,500 m^2 of mirrors. The *Solar-Two* upgrade adds another 11,500 m^2.

Dagan: What is the present cost of heliostats?

Klimas: We estimate that for the first fully commercial plant the cost of glass heliostats would be about US\$ 180 m^{-2} in today's dollars, or about US\$ 120 m^{-2} if enough heliostats were built for three plants. These are installed costs.

Dagan: What about stretched-membrane heliostats?

Klimas: All DOE work on this kind of technology stopped some time ago. We hope that the few that we shall install as part of the *Solar-Two* field upgrade will provide a boost to this kind of research.

Faiman: I have two questions about the economics of *Solar-Two*. First: On what basis is your figure of 5–10 US¢/kWh calculated; is it based purely upon the cost of the additional parts needed to upgrade *Solar-One*, or does it include an estimate of what the entire capital cost would be for this project were all the parts to be ordered now?

Klimas: The 5–10 US¢/kWh figure is only indirectly related to *Solar-Two*. It represents what we think the cost will be for 100–200 MW power plants. I'd hate to say what the cost will be for this one-of-a-kind research scale project.

Faiman: Second: I notice that the O&M cost you tabulate for *Solar-Two* is something like 10% of the capital cost, and yet in your projection to the 100–200 MW plants this figure goes down to about 1%. Can you indicate how such an order-of-magnitude cost reduction will be achieved?

Klimas: 10 MW is not an economic size for a solar power station. *Solar-One* demonstrated that fact. Even in the case of the Luz stations the O&M costs decreased as they increased the station size. At Harper Lake Luz were considering employing centralized control for four 80 MW plants. So, in a word, the answer to your question is economy-of-scale.

Gordon: Since the time of *Solar-One* there have been two major advances in the conceptual design of central receiver plants. One is the possibility of taking advantage of the high temperatures in order to employ gas turbines as part of a combined cycle system, rather than the thermodynamically inferior single steam turbine. Second is the desirability of employing second-stage concentrators in order to improve the optical efficiency. I am surprised that neither of these advances has been incorporated into the design of *Solar-Two*.

Klimas: The problem was that *Solar-Two* needed to be sold as a relatively low-cost, low-risk project. To this end it was necessary to make as few changes as possible in the original design.

Karni: But does your program also support research into more advanced ideas or is it doomed always to look only for near-term applications?

Klimas: For the last few years, unfortunately, survival has meant that we had to drop all of the more long-term research projects, even though these were our specialty. We hope that as our funding increases we will be able to return to such research but for the present time we can only undertake projects that appear attractive to the commercial sector.

Tabor: I am reminded, in this respect, of a comment that someone made at a conference in Canada: The problem of finding funding for solar research is that it has no military applications! This is a fact of life that solar people have to live with and it was one of the clever strokes of Luz that right from the start they conceived a system of a size for which funding could be raised from industry. Had they elected to go for a larger size system they would never have found the necessary financing.

Grasse: Between *Solar-Two* and the 100 MW scale how do you propose to test various technological improvements that may be expected to occur, such as stretched-membrane heliostats or advanced receivers, both of which are currently under test at Almeria?

Klimas: *Solar-Two* could easily become a *Solar-Three* test facility, by the incorporation of various advances such as those of the type you envisage. But we do not plan to build any plant intermediate in size between *Solar-Two* and the 100 MW plant.

Yogev: To what extent is a 100 MW plant part of the existing program or merely a dream at the present time?

Klimas: There are several commercial entities in the USA who regard this project as the first stage of a money-making enterprise. I believe that the fact that they are spending money on further studies is a positive indication that a 100 MW plant is a realizable goal.

Kearney: Luz had some mixed experience with regard to taking steps forward with advanced technology. For example, in going from a 30 MW to an 80 MW plant few questions were raised by the potential customers. However, when the selective surface was upgraded from black chrome to Cermet the most rigorous studies were required before they obtained the go-ahead. What would you say was the most critical component of *Solar-Two* in this regard?

Klimas: This, of course, is an illustration of why the *Solar-Two* design had to be so conservative with respect to innovations. Probably the most critical component is the new receiver, which has been completely redesigned.

Karni: Did you consider the possibility of employing *Solar-Two* as part of a hybrid system?

Klimas: This possibility was considered but dropped on political grounds: It was deemed to be a stronger selling point that the system should be nonpolluting.

Wolf: Why did you choose 4 hours of storage rather than some other figure?

Klimas: The 4 hours figure was a compromise between a smaller number that may have been optimal from the view-point of maximizing revenues from the summer afternoon demand curve peak, but would not have demonstrated, in such a clear manner, the "dispatchability" of solar energy that molten salt storage allows.

Saggie: Does the molten salt solidify during the hours when it is not used for storage?

Klimas: No. Calculations indicate that the insulated tanks are capable of maintaining the salt in its liquid state for several days.

Grasse: The Luz experience has taught us that it is necessary to have heavy involvement of big industry if solar is to succeed. Has US industry indicated any readiness to take over if *Solar-Two* is a success?

Klimas: The industry is already heavily involved. Companies such as Rockwell, Bechtel, Black & Veech and Science Applications International are already included and several other companies are negotiating to join.

Chapter 6

1997

6.1 Editor's Foreword

This chapter reproduces one of the two keynote lectures from what was actually the 8th Sede Boqer Symposium on Solar Electricity Production. The fact that contributions from the 6th and 7th symposia have been omitted from this volume is a reflection upon how the interest in solar power production has oscillated over the years, between photovoltaic and solar-thermal technologies. In particular, there were no solar-thermal key-note addresses during the two previous symposia. Fortunately, for the 8th symposium, Tom Mancini of Sandia National Laboratories was able to redress the balance, by focusing attention on the potential of parabolic dishes for solar-thermal power production. The omitted photovoltaic keynote presentations from the previous two symposia, together the lecture on concentrator photovoltaics that was given by Antonio Luque at the present symposium will appear in volume 2 of this series.

6.2 Dish Engine Systems: State of the Art, Challenges, and Opportunities (Dr. Thomas Mancini)

A keynote lecture presented by Dr. Thomas R. Mancini, Sandia National Laboratories*, Albuquerque, NM 87185, USA. [*Sandia is a multiprogram laboratory operated by Sandia Corporation, a Lockheed Martin Company, for the United States Department of Energy, under contract DE-AC04-94AL85000.]

197

6.2.1   *Introduction*

Of the three types of solar thermal electric power generation systems, trough electric, power tower, and dish engine systems, the dish/engine system has demonstrated the highest solar-to-electric conversion efficiencies, peak values of 29.4% and "annual average" efficiencies of 22%. A dish/engine system comprises a parabolic concentrator or dish, a thermal receiver, and a heat engine/generator. The system operates by tracking the sun and reflecting the solar energy to the focus of the dish where it is absorbed in the receiver. The heat absorbed by the receiver is then transferred to the heater head of an externally-fired engine/generator. The typical engine used in these systems is the Stirling engine (both kinematic and free-piston engines), although Brayton-cycle engines are planned for some systems under development. Dish/engine systems range in size from about 2 to $50\,kW_e$, primarily due to the sizes of available engines but also because of wind load limitations on large dish collectors. These units are modular, which allows their assembly into plants ranging in size from a few hundred kilowatts to tens of MW_e and should favor their early deployment in commercial power plants.

In this paper, I will describe the characteristics of the three major components of dish engine systems, present information on seven dish/engine systems, and briefly discuss market opportunities, costs, and barriers to the deployment of dish/engine systems.

6.2.2   *Characteristics of system components*

The three major components of a dish/engine system are the solar concentrator, the thermal receiver, and the heat engine/generator. Some of the characteristics and examples of different components that have been used in these systems are presented in this section of the paper.

6.2.2.1   *Solar concentrator*

The "ideal" solar concentrator is a parabola of revolution. In practice, the optical surface of the ideal concentrator is approximated with a series of facets, small or large optical surfaces that are intended to

Figure 6.2.1: Faceted test bed concentrator at Sandia National Laboratories (L) and the Schlaich Bergermann und Partner (SBP) continuous dish concentrator (R) [Photos courtesy Sandia Labs and SBP, respectively.]

model the performance of the surface of the paraboloid. Examples of a faceted and a continuous dish are shown in Fig. 6.2.1.

Even the "ideal" solar concentrator would not produce a point focus when imaging the sun, because the sun has a finite size equal to about 5.8 milliradians. In general, lack of ideality in solar concentrators arises from three areas of increasing scale: *specularity*, *slope error*, and *alignment*.

- Specularity is a measure of the smoothness of the surface and the uniformity of the silver coating layer. As such, it is related to the very small or microscopic scale of the surface.
- Slope error is a measure of the local deviation of the slope of the surface from that of a parabola of revolution. Its scale is larger than that of specularity, perhaps on the order of a centimeter.
- The third error parameter is the alignment of the facet or the general direction of the normal at its center point. Alignment can easily be off by several centimeters. It is not practical to measure these error parameters and, pragmatically, all three are generally "rolled up" into an equivalent slope error that is representative of the concentrator as a whole. Typically, we refer to a dish in terms of its "equivalent slope error," a one-standard-deviation

Optical Efficiency

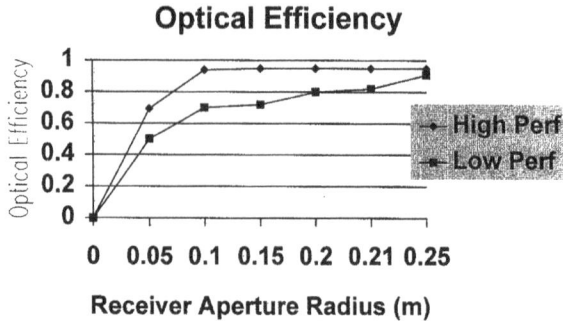

Figure 6.2.2:   Optical efficiencies of a high performance solar concentrator and a lower performance dish

value determined from a set of local measurements all assumed to be independent.

The equivalent slope error is shown graphically in Fig. 6.2.2 where the optical efficiency of two solar concentrators is plotted as a function of the receiver aperture radius. The optical efficiency of a dish is a function of the reflectivity of the surface and the intercept of the reflected sunlight in the plane of the receiver. The optical efficiency is defined as:

$$\eta_O = \text{(energy delivered to receiver aperture)}/$$

$$\text{(energy incident on the collector aperture)}$$

6.2.2.2   *Thermal receivers*

The key technology development issues of dish/engine systems are the reliable operation of the receiver under high-temperature and high heat-flux conditions, and the long life, high-efficiency operation of the engines. Heat fluxes of 50 to $75\,\mathrm{W/cm^2}$ and temperatures of 700 to $800°\mathrm{C}$ are common in thermal receivers, which are of two basic designs — direct illumination and heat pipes.

In the *direct illumination receiver* (DIR), the heater head of the engine is the thermal receiver. The engine working fluid, typically helium or hydrogen, is circulated through tubes on the engine head where it is heated by the concentrated sunlight. A typical DIR receiver is shown in Fig. 6.2.3.

Figure 6.2.3:   A typical direct illumination receiver [Photo Courtesy SBP]

Heat-pipe receivers use sodium or a mixture of sodium and potassium to transfer heat from the surface of the receiver to the engine heater head [Andraka, 1993 and 1995]. Heat pipes utilize a capillary wick to distribute the liquid metal over the back surface of the absorber. The liquid metal boils, the vapor is transported to the engine heater head where it condenses, and the liquid metal refluxes to the base of the absorber. Even though heat-pipe receivers only use about 1 kilogram of liquid metal, these materials are hazardous and handling, storing, and shipping receivers can be complicated. A schematic heat-pipe receiver that was proposed by Sandia National Labs is shown in Fig. 6.2.4.

Volumetric receivers have also been developed for potential use with Brayton-cycle engines. Most notably, the German DLR has developed a volumetric receiver for a Brayton engine that will be tested on sun at Sandia next year. In addition, the Weizmann Institute of Science has developed a "porcupine" receiver that could potentially be adapted to a Brayton engine. Volumetric receivers must have windows to maintain a pressure envelope and, consequently, can achieve higher temperatures and performance than DIRs or heat-pipe receivers. They are also likely to be of higher cost.

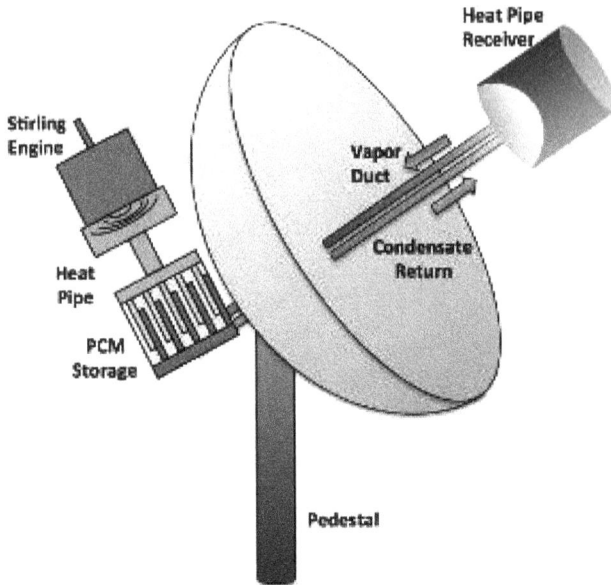

Figure 6.2.4:   Sandia heat-pipe receiver (schematic) [Figure courtesy Landia Labs.]

The three main heat loss mechanisms from a thermal receiver are conduction from the receiver, convection to the air, and radiation to the surroundings.

- *Conduction* is small, not often exceeding 10% of the total losses and more generally being on the order of 5%. The receiver is designed to minimize this loss.
- *Convective losses* can be larger and vary depending on the orientation of the cavity and the wind speed and direction. They are not readily modeled because of the lack of a consistent length scale and generally they range from 15 to 30% of the losses from the receiver.
- *Radiation* from the cavity to the environment is by far the largest thermal loss, normally representing 60 to 80% of the losses. Radiation losses are minimized by trying to operate the receiver at the smallest aperture possible. Of course, there is a tradeoff with the solar concentrator, which would like to operate at a relatively large aperture in order to maximize solar intercept.

6.2.2.3 *Heat engines*

Kinematic and free-piston Stirling engines represent the majority of engines used in dish/engine power systems. The kinematic engine utilizes the expansion and compression of the gas to drive pistons, and the power is removed through mechanical linkage to a generator. A free-piston engine has only two moving parts — the power and displacer pistons. Power is most often removed through a linear-alternator involving the displacer piston in a permanent-magnet-and-coil arrangement [West, 1986]. The technology status of kinematic Stirling engines is more advanced than that of free-piston engines. The tradeoff between the two engines is the simplicity of the free-piston engine versus the higher efficiency of the kinematic Stirling engine (30% for the free-piston versus 40% for the kinematic engine). A kinematic and a free-piston engine are shown in Fig. 6.2.5.

Two Brayton engines are also under development for dish/engine systems. Northern Research and Engineering Corporation of Nashua, New Hampshire, USA, is developing a Brayton engine from turbo charger parts. Allied Signal Corporation of Phoenix, Arizona, USA, is adapting a gas-fired turbogenerator set for on-sun testing at Sandia National Laboratories in April 1998.

6.2.2.3.1 Dish/Stirling systems

Over the last fifteen years, seven different dish/Stirling systems ranging in size from 2 to 50 kWe have been built by companies in

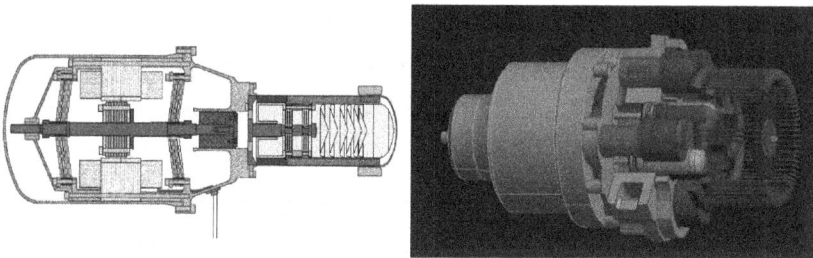

Figure 6.2.5: Schematics of the Sunpower free piston Stirling engine (left) and the Stirling Motors STM 4-120 kinematic Stirling engine (right) [Figures courtesy Sunpower and Stirling Motors, respectively.]

the United States and Germany [Stine, 1993 and 1994]. The first of these systems, the 25 kW$_e$ Vanguard system built by ADVANCO in Southern California, U.S., achieved a reported *World Record* net solar-to-electric conversion efficiency of 29.4% [EPRI, 1986]. Two 50 kW$_e$ dish/Stirling systems were built, installed, and operated in Riyadh, Saudi Arabia, in 1984. Details on these two systems are presented below.

6.2.2.3.2   Advanco's vanguard system

The 25 kWe. Vanguard system, built by ADVANCO in Southern California (1982–1985) is shown in Fig. 6.2.6. The Vanguard dish/Stirling system used a glass-faceted dish 10.5 meters in diameter, a direct insolation receiver (DIR), and a United Stirling 4-95 Mark II kinematic Stirling engine.

Figure 6.2.6:   The Vanguard dish/Stirling system [Photo courtesy Vanguard Inc.]

During the 18-month operation of the system, two engine over-hauls were required for two different reasons. The first overhaul was required due to the failure of a check valve, and the second overhaul was because of a failed oil pump shaft. Other problems, such as excessive noise, vibration, and repeated failure of circuit boards, seem to have been the result of using non-hardened gears. Also, there were substantial hydrogen leaks in the system necessitating the replacement of approximately $0.1\,m^3$ per hour of system operation. In spite of these problems, the system achieved the World's Record net solar-to-electric conversion efficiency of 29.4% [EPRI, 1986] and operated for almost 2000 hours during the 18-month test phase.

6.2.2.3.3 Schlaich-Bergermann und partner 50 kW system

Two 50 kWe dish/Stirling systems were built, installed, and operated in Riyadh, Saudi Arabia, in 1984 by Schlaich-Bergermann und Partner (SBP) of Stuttgart, Germany [Koshaim, 1986]. The dishes, shown in Fig. 6.2.7, were 17-meter diameter, stretched-membrane concentrators, formed by drawing a vacuum in the plenum space formed by the dish rim and front and back, 0.5-mm-thick stainless steel membranes. The optical surface of the dish was made by bonding glass tiles to the front membrane. The receivers for the SBP dishes were DIRs and the engines were United Stirling 4-275 kinematic Stirling engines.

6.2.2.3.4 Cummins power generation

From the late 1990s until 1996, Cummins Power Generation was developing two dish/Stirling systems as part of 50:50 cost-shared projects with the U.S. Department of Energy. In June of 1996, the Cummins Engine Company announced plans to "focus on their core capabilities" and, as a result, divested of their dish/Stirling system development. System hardware was sold to a Turkish holding company.

The smaller and more developed of the two Cummins systems was the 7 kWe system. This dish/Stirling system comprises a solar concentrator, a heat-pipe thermal receiver developed by Thermacore, and a free-piston Stirling engine by Clever Fellows Innovative

Figure 6.2.7: SBP 17-meter Dish/Stirling systems in Riyadh, Saudi Arabia [Photo courtesy SBP]

Consortium (CFIC). The solar concentrator utilizes a geodesic space frame, a polar-axis drive, and 24 five-foot-diameter, stretched-membrane, polymer mirror facets. The concentrator design was relatively mature but, since the development of long-life, reflective polymer films has not progressed as anticipated, CPG was considering a new solar concentrator design with a glass optical surface. The twin-opposed, dual-engine configuration of the CFIC free-piston engine eliminated vibration problems that were experienced in an earlier single-cylinder model. However, the development of the CFIC engine did not progress as quickly as CPG had anticipated and, as of August 1995, the engine had only demonstrated a maximum output of $5.2\,kW_e$ at an efficiency of 22%.

Through a second DOE cost-shared program, CPG was also developing a $25\,kW_e$ dish/Stirling system for grid-connected applications. The 25-kW project started in 1994 and used a new, high performance, solar concentrator design with a continuous glass surface comprising a number of pie-shaped gores. The dish was designed to provide

Figure 6.2.8: The Cummins power generation dish/Stirling system [Photo courtesy Cummins Inc.]

about 120 kW of thermal power to the receiver. The system operated only for a few hours in the spring of 1996 with a scaled-up version of the Thermacore^{TM} heat-pipe receiver and an Aisen Seiki, Japan, kinematic Stirling engine. It produced about 22 kW$_e$ output power. The Cummins dish/Stirling system is shown in Fig. 6.2.8.

6.2.3 Commercial development of dish/Stirling systems

Three companies, Stirling Energy Systems (SES) and Science Applications International Corporation (SAIC) in the United States and

Schlaich-Bergermann und Partner (SBP) in Germany, are currently developing commercial dish/Stirling systems. The technical viability of dish/Stirling technology for generating electrical power has been successfully demonstrated. Commercial development activities target the technical challenges of developing long-life systems and of making systems capable of producing competitively-priced electrical power.

6.2.3.1   *McDonnell Douglas aerospace dish/Stirling system*

In the early 1980s McDonnell Douglas Aerospace Corporation (MDAQ) developed a dish/Stirling system based on the Vanguard dish. When oil prices fell in 1986, MDAC discontinued development of the technology and sold the rights to the system to the Southern California Edison Company (SCE) [Lopez, 1992 and 1993]. More recently, the rights to the system were again sold to Stirling Energy Systems (SES) of Phoenix, AZ, USA. SES is upgrading and developing the system, shown in Fig. 6.2.1 (left), for commercial sales.

Operation during the mid 1980s demonstrated:

- 13,852 hours of operation of a single system (more than 19.4 operational years were accumulated on seven different systems);
- 86.5% system availability during the last two months of operation;
- Net power production when the direct-normal insolation exceeded 300 W/m^2;
- production of 28 kW net electric power at a solar irradiance of 850 W/m^2; and
- total production in excess of 720 kWh of net electrical power per day on a good solar day.

Based on their experimental measurements of the dish/Stirling unit, Lopez and Stone developed a predictive model of the system. Their results indicated that a single system operating at Barstow, California, U.S., would produce 50,122 kWh per year at annual efficiency of 22% (an annual availability of 90% was used in their analysis).

SES is currently upgrading the Kockums engine and marketing the system in the Southwest U.S.

6.2.3.2 *SAIC/STM system*

The team of Science Applications International Corporation (SAIC) and Stirling Thermal Motors (STM) has developed a 20 kW$_e$ dish/ Stirling system design that utilizes a faceted, stretched-membrane dish, a direct illumination receiver, and Stirling Thermal Motors' kinematic Stirling engine [Beninga, 1995]. This project, which is also a DOE cost-shared program, started in 1993. The Phase I system, shown in Fig. 6.2.9, produced 18.5 kW$_e$ during steady operation, delivered a peak power of 22 kW$_e$, and has accumulated more than 450 hours of on-sun operation. The team has redesigned the system to incorporate a larger dish that is also more easily manufactured, a hybrid DIR receiver capable of operating on natural gas or solar energy, and the next-generation of the STM 4-120 Stirling engine. This system is scheduled to operate on sun for the first time in December of 1997. During Phase II of the project in 1998, SAIC will install five systems at several locations throughout the U.S.

Figure 6.2.9: The SAIC/STM Phase I dish/Stirling system [Photo courtesy SAIC]

The Phase II dish was designed to accommodate mass production and ease of assembly and installation. It comprises a number of trusses, which are each made from identical members, and six identical radial trusses. Sixteen 3.0 m diameter facets are used on the dish resulting in a total area of 118 m². The facets are stretched membranes, two stainless steel membranes welded to a circular ring with a vacuum provide a plenum space to focus the facets. The optical surface of the facets is made by bonding thin glass mirrors to the surface of the facets.

The engine for SAIC's dish/Stirling system is the STM 4-120 kinematic Stirling engine, which was developed by Stirling Thermal Motors, of Ann Arbor, MI. The STM 4-120 Stirling engine is a four cylinder, variable-stroke, kinematic Stirling engine. The variable stroke is achieved through the use of a variable-angle swashplate. The system has operated on both hydrogen and helium and, since it provides a slightly higher performance with hydrogen, utilizes it in the generation two system.

The Phase I and II system performance, measured and predicted, are shown in Fig. 6.2.10.

Figure 6.2.11 shows a "waterfall" diagram representing the various efficiency losses from solar input to electrical output.

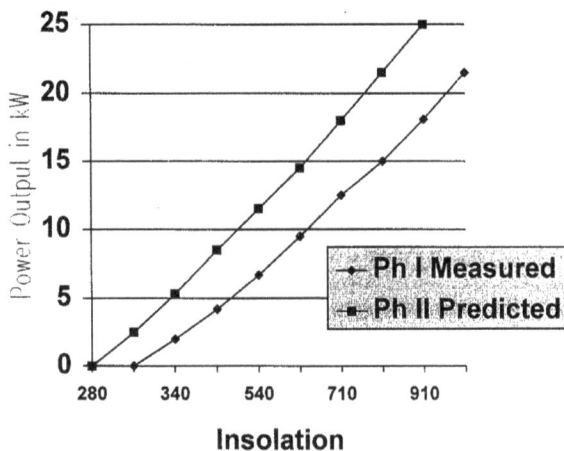

Figure 6.2.10:   Measured and predicted performance of the SAIC system

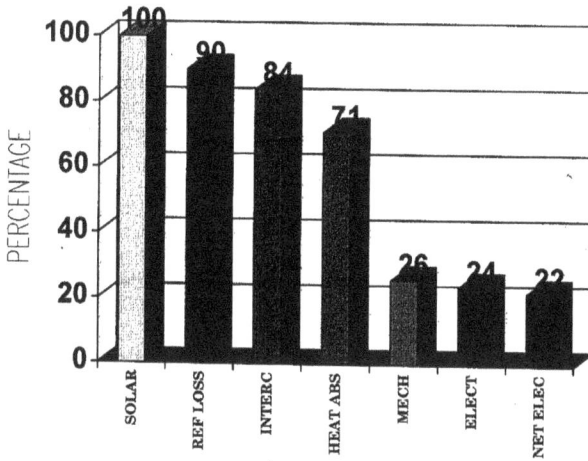

Figure 6.2.11: "Waterfall" Diagram for the SAIC dish/Stirling system

6.2.3.3 *Schleich-Bergermann und Partner (SBP)*

Schlaich-Bergermann und Partner (SBP) of Stuttgart, Germany has developed a 10 kWe system, shown in Fig. 6.2.12, that is the most-advanced dish/Stirling system available today and is nearing commercialization [Schiel, 1994]. The system is made of a 8.5-meter diameter, stretched-membrane concentrator, a direct-insolation receiver, and a V-161 kinematic Stirling engine manufactured by SOLO Kleinmotoren GmbH of Stindelfingen, Germany. Three systems have been on test since May of 1992 at the Plataforma Solar in Almeria (PSA), Spain [Sanchez, 1994]. At a solar insolation of 900 W/m^2, the systems produce about 9.5 kWe at an efficiency of 21%. Three second-generation systems were installed at the PSA for additional testing and, as of this writing, the six systems have accumulated more than 27,000 hours of on-sun operation.

The SBP solar concentrator is an 8.5-meter diameter stretched-membrane dish with 0.23-mm thick stainless steel membranes stretched on either side of a 1.2-meter wide outer ring. After the two membranes are attached to the ring, the front membrane is plastically formed using a combination of vacuum and hydrostatic forces to form a paraboloidal shaped surface. The dish contour is maintained with a

Figure 6.2.12:   SOLO 10 kW Stirling engine used in SBP system [Photo courtesy SBP]

low-level vacuum in the plenum space between the two membranes, and thin glass mirrors are applied to the front membrane to form the optical surface of the dish. The dish is mounted in a polar-type drive system that carries the gravity and wind load forces to ground through a highly-efficient tracking structure.

The SOLO Stirling engine, shown in Fig. 6.2.12 has a nameplate power rating of $10\,kW_e$ at 1500 rpm and a maximum power rating of $15\,kW_e$ at 3600 rpm. The V-161 engine is configured in the alpha arrangement with separate, single-acting compression and expansion pistons. The working gas is helium at a maximum operating pressure of 15 MPa and the power output is controlled by varying the pressure in the engine. Solar heat is transferred to the engine through a DIR receiver that operates at a maximum tube temperature of $740°C$, resulting in a maximum working helium temperature of $650°C$.

6.2.4 *Challenges and opportunities*

Dish/Stirling power technologies continue to develop rapidly and the cost of power is dropping. There are a number of initiatives that could potentially have more impact in bringing solar power to the marketplace than technology development.

- Tax equalization issues (in the U.S.) are being considered for a number of technologies [Jenkins, 1995]. Solar plants require a large, initial capital investment which is in a large part the purchase of a lifetime of fuel for the plant in the form of the collectors and receivers. Fossil-fired plants also pay property tax on their capital equipment (small compared with solar plants) but pay no taxes on their fuel. Consequently, proponents of solar plants feel that the capital equipment that provides the fuel should not be taxed in the same way as the power block.
- A second issue is the global concern for the environment and the quality of life. Since fossil plants emit large amounts of carbon dioxide and solar plants do not, some countries are considering ways to introduce carbon taxes into the decision-making process for new power plants. At the upcoming meeting in Kyoto, it is expected that most of the world's nations will respond by putting in place a carbon tax, although it is not expected that the U.S. will do so.

The current cost of dish/Stirling systems is in the range of US\$15–20/kW for prototype systems, US\$6/kW for production rates of 100s per year, and US\$2–3/kW at production rates of 1000s per year. The long-term target for costs at high production rates is in the range of US\$1.5–2/kW. These system costs translate into levelized energy costs for the systems in the near-term of US\$0.15–0.30/kWh. One of the important features of these systems is that they are readily hybridized. Hybridization allows these systems to provide base load, dispatchable power, making them more valuable in energy markets.

6.2.5 *Markets*

The first markets for these systems in the U.S. will likely be specialized, set-aside markets in the Southwest in Nevada, Arizona, and California. The emergence of "green power" markets may provide additional growth for the market once the existing resources have been allocated.

Mid-term markets will likely be for export to provide remote power and water pumping. Later markets may include small village electrification, end-of-line power conditioning, and eventually grid-connected power. The modularity of dish/Stirling systems limits the financial risks, since a small plant having all of the characteristics of a large one can be built and evaluated for a relatively small investment. The World Bank's Global Environmental Facility is currently discussing the support of projects in India, Egypt, Jordan, and countries in South America.

Some of the near-term challenges that we will have to overcome to deploy these systems are:

- Operation and maintenance data on existing systems will have to be developed further. Project developers have indicated a need for 5 years of good operation of a system before they will consider taking financial risks associated with deploying a new technology.
- Engine costs will have to decline substantially in order to achieve even the mid-term cost goals. The sale of dish/engine systems will not be sufficient to pace this cost decline, and we must rely on other markets to increase the production of the engines.
- The restructuring and re-regulation of global energy markets is causing the roles of the utilities, developers, and technology providers to change. In the U.S., large investor-owned utilities will no longer generate power.
- Who will be willing to take the technology risk associated with deploying a new technology such as this one? IPPs?
- Today's energy project investment is made in the no-risk technology that provides the least expensive power.
- Long-term markets must demonstrate that they are sustainable in order to allow manufacturers to invest in infrastructure and plan for the future.

6.2.6 *References*

Andraka, C., D. Adkins, T. Moss, H. Cole and N. Andreas. "Felt-Metal-Wick Heat-Pipe Receiver" *Solar Engineering 1995*, Proceedings of the ASME/JSME/JSES International Solar Energy Conference, Maui, HI, March 1995, ISBN 0-7918-1300-2.

Andraka, C., R. Diver, D. Adkins, S. Rawlinson, P. Cordeiro, V. Dudley and T. Moss. "Testing of Stirling Engine Solar Reflux Heat-Pipe Receivers" Proceeding of the 28th Intersociety Energy Conversion Conference (IECEC), Atlanta, GA, August 1993, ISBN 0-8412-2772-5.

Bean, J.R. and R.B. Diver "Performance of the CPG 7.5 kW$_e$ Dish/Stirling System" Proceedings of the 28th IECEC, Atlanta, GA., August 8–13, 1993, pp. 2.627–2.632, ISBN 0 7918-1221-9.

Bean, J.R. and R.B. Diver. "Technical Status of the Dish/Stirling Joint Venture Program" Proceedings of the 30th IECEC, Orlando, FL, August 1995, pp. 2.497–2.504.

Beninga, K. and R. Davenport. "Design, Testing, and Commercialization Plans for the SAIC/STM 20 kW$_e$ Solar Dish/Stirling System" Proceedings of the 30th IECEC, Orlando, FL. August 1995, pp. 2.487–2.490.

EPRI Report "Performance of the Vanguard Solar Dish/Stirling Engine Module" Electric Power Research Institute, AP 4608, Project 2003-5, July 1986.

Gallup, D.R., and T.R. Mancini, "The Utility-Scale Joint-Venture Program," Proceedings of the 29th IECEC Conference, Monterey, CA, August 7–12, 1994. ISBN 1-56347–91-8.

Goldberg, V., J. Ford and J. Anderson. "Design of the Support Structure, Drive Pedestal, and Controls for a Solar Concentrator" Sandia National Laboratories Report, Albuquerque, NM, SAND91-7007, August 1991.

Grossman, J., R. Houser and W. Erdman. "Testing of the Prototype Facets for the Stretched-Membrane Faceted Dish" Sandia National Laboratories Report, Albuquerque, NM, SAND91-2202, December 1991.

Lopez, C. and K. Stone "Design and Performance of the Southern California Edison Stirling Dish" Solar Engineering, Proceedings of the ASME International Solar Energy Conference, Maui, HI, April 5–9, 1992, ISBN 0-7918–762-2.

Lopez, C. and K. Stone. "Performance of the Southern California Edison Company Stirling Dish", SAND93-7098, Sandia National Laboratories, Albuquerque, NM, October, 1993.

Mancini, T.R. "Analysis and Design of Two Stretched-Membrane Parabolic Dish Concentrators," *ASME Journal of Solar Energy Engineering*, August 1991.

Mancini, T. R. "An Overview of Concentrators and Receivers for Solar Thermal Systems," invited keynote presentation at the 7th International Symposium on Solar Thermal Concentrating Technologies," Moscow, Russia, September 26–30, 1994.

Mancini, T.R. "An Overview of Parabolic Dish Concentrator Development," Proceedings of the Fifth Task III Meeting, Solar PACES, Paul Scherrer Institute, Villigen, Switzerland, March 1995.

Mancini, T.R., J.M. Chavez and G. J. Kolb. "The Promise and Progress of Solar Thermal Power," *Mechanical Engineering Magazine*, Vol. **116**, No. 8, August, 1994.

Mancini, T. R., G.J. Kolb and M.R. Prairie. "Advances in Solar Energy", Chapter on Solar Thermal Power, publication of the American Solar Energy Society, Spring 1997

Sandubrae, J., K. Beninga and J. Halford. "Structural Analysis and Design of a Faceted Stretched Membrane Dish" Solar Engineering 1995, Proceedings of the ASME/JSME/JSES International Solar Energy Conference, pp. 539–544, March 1995, ISBN 0-7918-1300-2.

Schiel, W., T. Keck, J. Kern and A. Schweitzer. "Long Term Testing of Three 9 kW Dish/Stirling Systems" ASME International Solar Energy Conference, San Francisco, CA, USA, March 1994.

Stine, W. and R. Diver. "A Compendium of Solar Dish/Stirling Technology" SAND 93-7026, Sandia National Laboratories, Albuquerque, NM, January 1994.

Stine, W. "An International Survey of Parabolic Dish/Stirling Engine Electrical Power Generation Technology" Solar Engineering, Proceedings of the ASME/ASES Joint Solar Energy Conference, April 4–8, 1993, Washington, D.C.

U.S. DOE, Solar Thermal Electric Five-Year Program Plan FY 1993 through 1997, Solar Thermal and Biomass Power Division, Office of Solar Energy Conversion, U.S. Department of Energy, Washington, DC.

West, C. Principles and Applications of Stirling Engines, Van Norstrand Reinhold, New York, N.Y., 1986, ISBN 0-442-29273-2.

6.3 Discussion Following Tom Mancini's Presentation

Luque: At your location how many kWh per year would you get out of each kW of system?

Mancini: Around 2,500–3,000 kWh per year.

Question: When you mentioned US$20/W for the dish system how much was engine and how much dish?

Mancini: About 50–50, or US$0.25M for each in the prototype.

Gordon: Why do you not employ optical means, such as a kaleidoscopic homogenizer, to overcome the problem of hot spots?

Mancini: It's an interesting idea but one that requires much more detailed study. For example, placing mirrors in an 800° environment may be problematical for the mirrors and the resulting change in cavity geometry may change the radiation balance.

Medwed: I have been told by a couple of suppliers with whom I am negotiating a fairly large PV project that, because of the German "thousand roof" and the US "million roofs" initiatives, supply cannot meet the demand and PV prices are actually rising. Doesn't this seem an odd sort of incentive?

Mancini: I totally agree but there are also other ways in which these government initiatives harm the technologies they seek to promote. I recently had a conversation with an attorney who works for the Arizona Public Utility. He confessed that the million roof project scares him for another reason: all of the law suits he believes will follow in its trail as soon as the PV systems begin to malfunction. He recalls the situation in the 1970s when vast quantities of highly subsidized but poorly designed and installed solar water heaters lead to leaky roofs. In his opinion the ensuing legal situation was one of the major contributory factors to the death of solar water heating.

So yes, indeed: large demonstration projects can do more harm than good to the technology they aim to support if they are not carefully thought through in advance.

Roy: What about wind loads? Are they not a problem?

Mancini: They are indeed and the only way I know of to overcome them is to make the dishes smaller.

Gordon: Instead of mounting a heavy engine, high in the air, at the focus of a large dish, why not use many small dishes and employ fiber optical means to bring the flux to a ground mounted engine? You might even think of employing secondary optics indoors to improve the efficiency.

Luque: We have given this idea some thought in the context of concentrator PV. We concluded that the large scale use of optical fibers would tend to increase costs.

Returning to the dish/Stirling system, can you give us a rough cost breakdown of the kind of system you would regard as being commercially viable?

Mancini: For the first market we might get into I could imagine a 25 kW system for US$6/W, i.e., US$150K. Of this, the engine would cost US$70–80K. The concentrator would cost a similar amount, of which US$20–30K would be the cost of the foundations. One of our major concerns is the drive unit. That is why Jeff Gordon's suggestion to use fiber optics is not without its appeal.

Roy: Many Third World countries have no backup fuel resources but they do have a surfeit of low-cost labor. How does this fact affect the potential market for solar-thermal systems?

Mancini: You have identified a major problem that we have been working on but we still have no clear answers. We have identified a number of components whose costs could be substantially reduced but whether we could come up with a system suitable for this kind of foreign market is still questionable.

Question: You are advocating a complex technology with an efficiency around 25%. On the other hand, PV has an efficiency of around 15% but is technologically much simpler. Would you care to comment?

Mancini: I prefer not to analyze the relative costs of the two technologies at the present state of the art. I believe we are at a stage where a break-through in either technology would be good for both.

Luque: I agree. Your technology, for example, has some very important advantages and some drawbacks. The most impressive advantage is the 25% efficiency but, in my opinion, the most serious drawback is the high cost of the components.

Roy: Two years ago, at this symposium, Petr Svoboda of Pilkington Solar argued that a 10% solar share in a 100 MW combined cycle Siemens plant would result in a cost benefit of some 5 US¢/kWh. Might this not be a direction to explore in order to keep dish/Stirling systems alive?

Mancini: Unfortunately not. Dish/Stirling systems, unlike simple solar fields which merely add heat to the entire system, are entire systems themselves. One could, of course, think of using a field of dishes in similar manner to the way troughs have been proposed but that is not a dish/Stirling system.

Chapter 7

2001

7.1 Editor's Foreword

This chapter skips over the 9th Sede Boqer Symposium, which was held in July 1999, because there were no solar-thermal keynote presentations at that meeting: only photovoltaics. By way of contrast, the 10th Symposium, held in February 2001, had one photovoltaic keynote, delivered by Frederic Leenders; another keynote, by Harald Ries, on a kind of inverse non-imaging problem, in which Dr. Ries showed how specially designed optics can be used to enable broadcasts from a satellite transmitter to be spread uniformly over an island of arbitrary shape without any spillage into the surrounding sea; and one genuine solar-thermal presentation, which is reproduced here. The omitted photovoltaic keynote presentations from the previous and the present symposium will appear, respectively, in volumes 2 and 3, of this series.

7.2 Solar Thermal Electricity: Where do We Go from Here? (Dr. David Mills)

Invited keynote presentation given by Dr. David R. Mills, Department of Applied Physics, University of Sydney, Sydney, Australia 2006. (The written manuscript delivered by Dr. Mills for publication in the Symposium Proceedings was co-authored with Chris J. Dey; Philipp Schramek, Technical University of Munich, München, Germany; and Graham L. Morrison, University of New South Wales, Sydney, Australia.)

7.2.1   *Abstract*

A low demand scenario, based upon industry growth rate assumptions suggests that direct solar electricity, because it is starting from a very low current industry size, is unlikely to contribute significantly to global emissions reduction before 2035, but becomes essential to meeting climate goals after this time. To fulfil this long-term role, rapid growth must ensue now and be continued for decades so that full market potential can be achieved. Given this need, in this paper we argue for a broadened new industry market approach, which has important differences from current Solar Thermal Electricity (STE) industry visions for the future.

The main features of the suggested industry approach outlined in the paper can be summarized as follows:

- The primary market for STE should mostly be in developed countries, rather than in developing countries under Global Environmental Fund (GEF) projects.
- Until 2035, two primary transition markets that use hybridization (but not thermal or chemical storage) are recommended to create strong STE industry growth. These applications can use low cost line focus technology. They are:

  1. Solar biomass hybrids with a significant solar fraction to provide firm capacity in both developed and developing countries. In developing countries which can use fossil fuel under climate agreements, solar fossil (mainly coal) hybrids could be used.
  2. Solar coal savers supplying reheat thermal energy or main boiler steam for developed nations at a low solar fraction.

The main bounds on future growth will come from the performance of renewable energy competitors rather than fossil fuel. Evidence presented in this paper suggests that solar thermal electricity and other renewable energy options are likely to be less expensive as a total societal cost than conventional fuel, and may be highly competitive against pure wind and biomass technology without the invocation of storage. Future storage options applied to line focus technology may improve avoidance cost of atmospheric carbon, as will large scale production.

This paper does not address policy issues, but it is essential to the development of the STE industry that full life cycle costing information be adequately developed as an essential energy policy input so that equitable societal costing of different energy options can be performed.

7.2.2 *Introduction*

In 1995 the *Intergovernmental Panel on Climate Change* (IPCC), a synthesis of many climate experts, recommended [1] that an immediate reduction of at least 50% of current carbon dioxide emissions is required to stabilize carbon concentration in the atmosphere. The deployment strategy for renewable energy technology will have a significant bearing on the total emissions reductions that are possible over the next 50 years.

Solar thermal and other concentrating solar technologies will play a big role in this effort. However, a broader and more aggressive market policy may be required.

There is much to learn from the examination of a long-term model with a variety of solar and renewable energy technologies jointly represented. In particular, my colleagues and I feel that it is important to investigate whether a plausible mix of familiar energy technology is likely to meet the requirement of atmospheric stabilization by 2050 under a highly energy efficient scenario. This can be based upon examination of the likely industry growth potential of each industry. This approach will also give a feel for the role direct solar and solar concentrating technologies might play.

Figure 7.2.1 shows a projection which was prepared by Mills and Dey [2] using information on expected conventional energy growth to 2015 and renewable industry growth in a number of sectors out to 2070. It assumes that energy efficiency is used to strongly reduce energy demand by 2050. The 2050 demand total is estimated as 614 EJ, the mean of the low demand estimations mentioned above. We assume in Fig. 7.2.1 that demand after 2050 is limited

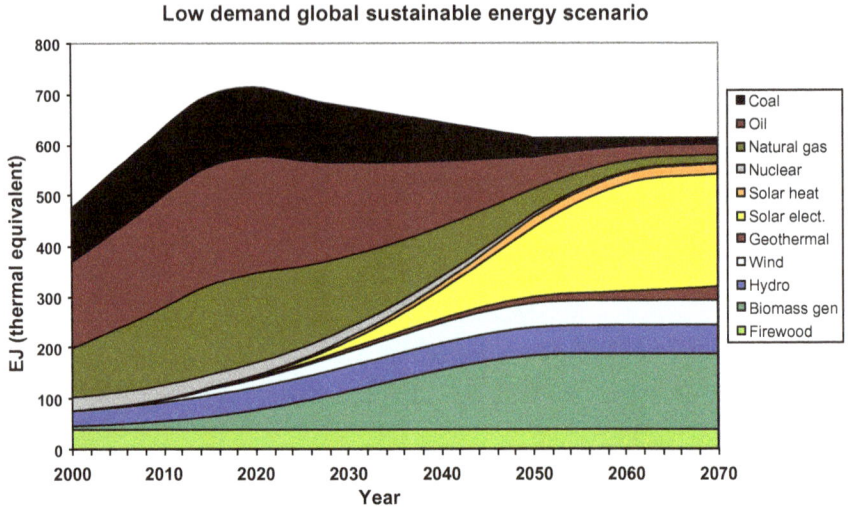

Figure 7.2.1:   Low demand scenario for fossil fuel replacement [2]. In this case, direct solar electricity installations are lumped together and cumulative capacity assumed to expand at almost 25% per year until 2035, when the annual capacity increase is frozen at the same level as today's natural gas increase

to 614 EJ. This is a simple assumption which reflects an expectation that changes will be relatively much reduced under the influence of marginally improving energy efficiency, smaller population changes, and a stabilizing per capita energy services demand.

In Fig. 7.2.1, the future global energy service demand is shown based upon U.S. Department of Energy, *International Energy Outlook* energy demand projections out to 2015 [3], but with slight adjustments near 2015 due to a larger assumed fraction of renewable energy in the mix and slight rounding of the peak near 2015 (any global change in demand is likely to be even less abrupt than this, because demand reduction is a gradual process; prediction of the likely demand shape is however, beyond the scope of this paper). Beyond 2015, a simple linear demand reduction between 2015 and 2050 is assumed, wherein the relative proportions of conventional fuels used in the IEO fuel mix of 2015 are maintained, and the total contribution of conventional fuel is simply reduced to the difference between the expected low demand load and the expected renewable energy supply based upon renewable energy industry expectations in that year.

Figure 7.2.1 is not an economic model, but an estimate of demand combined with an industry capability assessment. We do not speculate on the policy or economic measures required to decrease energy demand in this way or to increase the renewable energy market share to the levels shown. We merely assume that effective policy has been introduced. Details of the model assumptions can be found in Ref. [4]. However, in the scenario, limits are placed on the growth of the renewable industry such that the industry does not venture into growth rates that are outside current experience for fossil fuel. The low demand scenario for 2050 represents one of many possible scenarios, but it is felt to be a cost-effective scenario. A low demand scenario with renewable energy is probably a cheaper means of achieving atmospheric stability than a high demand scenario using even more renewable energy, and it is probably the cheapest of any scenario when total environmental costs are included.

The combined market for STE and PV today is tiny at about 0.02 EJ. PV has the technical capability to operate as a large electricity source, but must drop costs greatly in order to compete against present and future grid based STE. Nevertheless, in Fig. 7.2.1, no distinction is made between direct solar electricity contributions from photovoltaic (PV) and solar thermal electric (STE) installations. These are lumped together and cumulative direct solar electricity capacity is assumed to expand at almost 25% per year until 2040, ultimately being limited by market demand and reasonable industry growth rates rather than resource shortage. The majority of solar electricity capacity after 2035 would be expected by the authors to be grid based solar thermal electricity (STE) incorporating storage to allow 24 hour operation, or STE plants hybridized with biomass fuel.

Figure 7.2.1 shows the importance of biomass and solar electricity in meeting the 2050 goals. Direct solar energy starts from too small a base to be the leading fuel by 2035, but is comparable to combined firewood and biomass by 2050 exceeds them by 2070.

Summarizing this section, this low demand scenario, which is likely to be the cheapest for global society (given that high demand

scenarios involve substantial external environmental and social costs)
is characterized as follows:

- Direct solar electricity is unlikely to contribute significantly to
  global emissions reduction before 2035, but becomes essential to
  meeting climate goals by 2050.
- Direct solar energy starts from too small a base to be the leading
  fuel by 2050, but direct solar heat and electricity are approximately
  equal to combined firewood and biomass by 2070.
- Storage plant for wind and solar is unlikely to be required in large
  quantity until after 2035.
- Conventional fuel usage does not fall below present levels until
  after 2035.
- Fossil fuels must enter steep decline after 2015.

7.2.3   *Direct solar electricity generation*

As stated in the previous section, any commercial contribution
from direct solar electricity is starting from a very low base. Total
capacities of about $1\,GW_e$ of PV and $0.35\,GW_e$ of STE have been
installed, together covering about 1/6 of the total direct thermal
equivalent solar production, with virtually all of the remainder being
solar hot water capacity.

PV has the technical capability to operate as a large electricity
source, but technical means must be found to address high capacity
factor markets, and costs must be reduced greatly in order to compete
against grid based STE and wind technology [5, 6]. In contrast,
concentrating STE has both the potential for large scale deployment
and relatively low cost. It has the advantage of flexibility, being able
to interface with thermal and chemical storage systems and with
various backup fuels. The pressing immediate concern is whether
storage is needed. Does storage make STE any more attractive to
the anticipated transition market of the next two or three decades,
or should STE simply be hybridized with other fuels and achieve the
necessary industry growth goals? To answer this, we need to look at
the technologies involved.

7.2.4 *Current STE technology*

Up until now, parabolic trough technology [7, 8] has been the most commercially developed technology, as demonstrated by the Kramer Junction plants in California. Near term future estimates which resulted from a recent workshop in Colorado are that parabolic trough technology is the only type of STE technology that can be deployed immediately with acceptable financial risk in competitive power markets [8, 9, 10]. Cohen *et al.* [8] report that the full effect of improvements could realize a levelized electricity cost (LEC) of 0.04–0.05 US$/kWh in the long term.

Central receiver (solar tower) technology, typified by the Barstow Solar-One and Solar-Two plants, is a second basic collector system option. Solar-Two has demonstrated 24 hour solar operation using molten salt as the heat transfer and storage medium [11, 12]. Kolb [11] estimates that such an advanced central receiver system in solar only mode will have a levelized electricity cost of 0.065 US$/kWh$_e$ in high production.

Parabolic dish systems are a third broad type of STE technology. In these systems, a parabolic dish focuses light onto a thermal receiver which may be a Stirling engine or a steam generator. Steam may be generated at the receiver, or transported to a centralized turbine. These are usually targeted at remote markets, but some estimations of long term, high production cost are similar to that of the other technologies. The Australian National University has been involved in the development of large, potentially grid-connected steam generating dishes for some years but significant deployment has not yet occurred.

A fourth option being developed is linear Fresnel reflector technology, in which long reflector rows are aimed at an inverted linear absorber on a linear tower. My colleagues and I are involved in developing a highly ground area efficient form of this technology called the Compact Linear Fresnel Reflector (CLFR) for a commercial project in Queensland, Australia [13]. This incorporates saturated steam, direct steam generation (DSG), and allows fixed steam pipes and minimal reflector support structure.

7.2.5   *Direct costs of STE technology for proposed transition line focus options*

Direct costs are the main factor determining whether STE generation technologies are economically viable. Direct costs do not include societal and environmental costs and benefits. Currently the direct costs of STE are usually higher than for fossil fuel, but STE arrays can be manufactured as an industrial assembly line product and are subject to classical drops in price with increased production. The wind industry has illustrated this very well.

Kolb [11] has already provided an excellent description of the long term potential of central receiver options. In this section, I will attempt to use a broadly compatible economic analysis to examine whether line focus technologies, used in suggested market niches A and B (defined below in Section 7.2.7) in developed nations, might be more cost effective than 'GEF Towers' using storage systems.

For parabolic troughs, DSG developments are projected to lead to a 5% reduction in total system cost [7]. However, this should be compared to the estimated cost reductions quoted in the same reference for improved plant system design (10%), larger plant size (10–15%), multiple plant procurement (10–15%), standardized engineering (5%) and competitive pressures (15–20%). Thus, the present day cost (~$US0.10 per kWh) of stand-alone solar trough/gas hybrids could be dropped to approximately half, but increased market size and competition will have a far greater potential impact than solar plant technical improvements. In short, the establishment of large viable low cost transition markets with large-scale plants will have a profound effect upon solar thermal electricity costs, just as it already has had upon wind generation costs.

In Table 7.2.1, the assumptions behind the calculations shown in subsequent Tables 7.2.2 and 7.2.3 are presented. The cost multiplier for fuel and operation and maintenance (O&M) expenditure which I use is 1.45. Kolb [11] uses 1.5 for a 30 year lifetime, but this was adjusted in the present lecture to account for a 25 year lifetime. Calculations for troughs in Ref. [7] do not seem to include escalation in fuel and O&M costs. The fixed charge rate (FCR) in Ref. [7] is approximately 0.085 for the fossil capital fraction and about 0.1 for

Table 7.2.1:   Summary of assumptions for calculations

Lifetime [years]	25
Inflation	4%
Fixed charge rate for solar field capital	0.1
Fixed charge rate for fossil and biomass capital	0.1
Fuel cost and O&M cost multiplier [25 years]	1.45
Displaced capital cost for coal plant [US$/kW$_e$]	1140
Displaced capital cost for gas plant [US$/kW$_e$]	410
Displaced O&M cost for coal plant [US$/kWh]	0.0052
Displaced O&M cost for gas plant [US$/kWh]	0.0040
Capital cost for biomass plant [US$/kW$_e$]	2000
Solar/biomass O&M costs [US$/kWh]	0.006
Combustion efficiency of coal and biomass	90%
Combustion efficiency of gas	82%
CO_2 emissions content for coal [gCO_2/kWh]	909
CO_2 emissions content for gas [gCO_2/kWh]	554
CO_2 emissions content for biomass [gCO_2/kWh]	50–150

the solar capital fraction (given the financial parameters listed on page 68 of that reference). The FCRs used by Kolb [11] are 0.16 and 0.1, for fossil and solar capital respectively. My colleagues and I believe that this difference cannot be assumed, and have used an FCR of 0.1 for both solar and non-solar components.

Table 7.2.2 shows the projected present and future LEC of examples of coal saving CLFR and trough technologies compared to standard LS3 trough/gas plant. LS3 and CLFR plant performance is modelled on the basis of minute by minute solar radiation for the site of Rockhampton, as supplied by the Australian Bureau of Meteorology, with the measurement site close to the first CLFR plant being built. The Rockhampton coastal site has considerably less direct normal beam than US desert sites but this is compensated by a lower latitude location, where annual cosine losses are substantially smaller. Most inland Australian sites at this latitude would deliver substantially higher performance than US desert sites for this reason, but fewer coal offset opportunities exist in Australia than in the United States. Note that the installed cost of the CLFR solar capacity (US$/kW$_e$) appears to be particularly low compared to LS3 systems, in part because the CLFR has a high *peak* solar output.

Table 7.2.2: Examples of coal-saver trough and CLFR plants compared to a standard trough/gas hybrid. The initial solar base case is for a Nevada site [7] but the remainder are calculated for Rockhampton in north-eastern Australia. Assumptions are given in Table 7.2.1 and described in the text

Solar plant type Backup system	LS3 Gas	LS3 Gas	LS3 Gas	LS3 Coal boiler	CLFR Coal reheat	CLFR Coal reheat	CLFR Coal boiler
Location	Nevada	Rockh.	Rockh.	Rockh.	Rockh.	Rockh.	Rockh.
Time frame/production scale	1995	1995	Mod. prod.	Mod. prod.	First plant	Mod. prod.	Mod. prod.
Mirror aperture (m²)	470880	470880	470880	470880	17280	488100	488100
Circulating fluid temperature (¿C)	390	390	390	390	285	285	365
Net capacity (MWe)	80	80	80	80	4.08	115	143
Ave global insolation (MJ/m²/day)	—	20.0	20.0	20.0	20.0	20.0	20.0
Avg. daily solar output (MJ/m²)	11.5	10.8	11.3	11.3	11.0	11.0	10.7
Turbine efficiency	0.37	0.37	0.37	0.39	0.315	0.315	0.39
Equivalent full load hours (h/yr)	7008	7008	7008	2634	1487	1487	1442
Annual electric output (GWh/yr)	561	561	561	211	6.07	171	206
Solar share	0.36	0.34	0.36	1.00	1.000	1.00	1.00
Annual solar electric output (GWh/yr)	203	190	200	211	6.07	171	206
Direct costs (000s US$/yr)							
Site Works	7733	7733	included	included	included	included	included
Solar Field	101091	101091	included	included	included	included	included
HTF System/Boiler	21054	21054	included	included	included	included	included
Power Block	38037	38037	included	included	0	0	0
Balance of Plant	17395	17395	included	included	included	included	included
Land	498	498	included	included	10	277	277
Indirect Costs	34262	34262	included	included	included	included	included
Project total (000s US$)	220070	220070	160000	129371	4066	61589	64669
Unit Cost US$/kWe	**2751**	**2751**	**2000**	**1617**	**996**	**534**	**453**

(Continued)

Table 7.2.2: (*Continued*)

Fuel cost (US$/GJ)	2.17	2.32	2.32	0	0	0	0
Fuel cost (US$/kWh)	0.026	0.028	0.028	0	0	0	0
Fuel cost (000s US$/yr, $\eta_{Combustion} = 0.82$)	9184	10191	9929	0	0	0	0
Annual O&M cost (000s US$)	4764	4764	3573	2537	50	1059	1059
LEC (US$/MWh) — total	**75.3**	**77.9**	**63.5**	**78.8**	**78.9**	**44.9**	**38.9**
Capital cost fraction	39.3	39.3	28.5	61.4	67.0	35.9	31.4
Fuel cost fraction	23.8	26.4	25.7	0.0	0.0	0.0	0.0
O&M cost fraction	12.3	12.3	9.2	17.5	11.9	9.0	7.5
LEC (US$/MWh) — solar only	**111.7**	**119.2**	**83.1**	—	—	—	—
Capital cost fraction	87.6	93.4	64.7	—	—	—	—
O&M cost fraction	24.1	25.8	18.4	—	—	—	—
LEC US$/kWh — total	**0.075**	**0.078**	**0.063**	**0.079**	**0.079**	**0.045**	**0.039**
LEC US$/kWh — solar only	**0.112**	**0.119**	**0.083**	**0.079**	**0.079**	**0.045**	**0.039**
Yearly emissions avoidance cost							
Displaced fuel cost ($/kWh) — high	0.0300	0.0300	0.0300	0.0300	0.0300	0.0300	0.0300
Displaced fuel cost ($/kWh) — low	0.0025	0.0025	0.0025	0.0025	0.0025	0.0025	0.0025
Displaced capital cost ($/kWe)	1140	1140	1140	0.000	0.000	0.000	0.000
Displaced O&M cost ($/kWh)	0.0052	0.0052	0.0052	0.0052	0.0052	0.0052	0.0052
Total displaced cost ($/kWh) — high	0.067	0.067	0.067	0.051	0.051	0.051	0.051
Total displaced cost ($/kWh) — low	0.027	0.027	0.027	0.011	0.011	0.011	0.011
Net cost ($/kWh) — high	**0.008**	**0.011**	**−0.004**	**0.028**	**0.028**	**−0.006**	**−0.012**
Net cost ($/kWh) — low	**0.048**	**0.051**	**0.036**	**0.068**	**0.068**	**0.034**	**0.028**
Actual emissions (g CO_2/kWh)	554	554	554	0	0	0	0
Alternate emissions (g CO_2/kWh)	909	909	909	909	909	909	909
Emissions saved (t CO_2)	**324700**	**316100**	**322600**	**212900**	**6130**	**173200**	**208000**
Net cost/tonne CO_2(high fuel cost)	**$13.8**	**$18.8**	**−$6.7**	**$27.5**	**$27.6**	**−$6.1**	**−$12.1**
Net cost/tonne CO_2(low fuel cost)	**$82.7**	**$89.6**	**$62.6**	**$67.0**	**$67.1**	**$33.4**	**$27.4**

The overall 2000 US\$/kW$_e$ figure for new LS3 plants comes from Ref. [9]. The figure has been accepted without further examination as a large production figure using current technology, as the reference did not provide a detailed breakdown. I have scaled the fossil/solar components in trough collectors from the Pilkington report to give updated values for the capital component of the LEC. For CLFR plant, an initial commercial coal saver plant is costing A\$7 million for 4.2 MW$_e$, of which A\$2 million is supplied as a government grant for up front development, engineering and one-off startup costs. This plant is listed so, but a second, larger plant would cost approximately 5/7 of this in cost per peak kW$_e$, with no assumptions of significant equipment price reductions. In the table, a moderate production case is assumed with the CLFR in significant production, and a 25% price drop is assumed in cost per peak kW$_e$ over the second plant figure. This price drop is similar to that assumed by the trough industry for plant built today versus plant built in 1990. Long term high production would yield additional cost drops, possibly similar to the 50% estimated for trough technology, but these costs are not used here.

In addition to the LEC, the net cost increase per kWh over a base system is also an important quantity to be identified because it can be compared to net pollution gains. For stand-alone hybrid plants, my colleagues and I believe that the full cost of the capital and fuel of the cycle replaced must be subtracted from the cost of the total solar hybrid project and fuel cost to obtain the net cost against which emissions can be applied. However, in some cases this can be a mix of technologies on the electricity grid, such as coal, natural gas turbines, and natural gas combined cycle. If the technology can deliver a high capacity factor, however, it can be said to offset baseload coal. On the other hand, as stated earlier we believe that it is correct to look at retrofit fuel saver projects from the point of view that initial spending on the pre-existing fossil fuel plant is sunk capital. The retrofit solar project simply includes the life cycle cost of alterations to the fossil fuel plant less life cycle fuel costs saved on that site. Fuel costs should include the fuel retail costs plus fuel handling and waste dumping costs.

In Table 7.2.2, it can be seen that in terms of direct costs, solar options are more expensive than the assumed displaced direct coal generation costs ranging between 0.027 to 0.067 US\$/kWh$_e$. The net costs over coal generation range from 0 to 4 US¢/kWh$_e$ for an LS3/natural gas hybrid built today to -1 to $+3$ US¢/kWh$_e$ for a future CLFR coal saver. The retrofit coal savers offer a better net LEC and have the advantage that projects can be built quickly because they use existing generating facilities and grid infrastructure. The high and low fuel cost cases for the displaced electricity include extremes of 3US¢/kWh$_e$ down to 0.25US¢/kWh$_e$. At present, Australian coal costs are well towards the lower end of this range.

ISCCS (Integrated Solar Combined Cycle System), is not costed here because there is insufficient openly published cost data on the cycle. However, the operation of ISCCS is such that roughly half of the bottom cycle energy is supplied by solar at peak.

When solar is not available, the CC plant operates at reduced capacity such that the waste heat from the gas turbine supplies all of the energy to the steam turbine. Because of this, the solar output may be regarded as an uncontrolled renewable energy generator replacing not gas on the site, but a mixture of other fuels on the grid. The marginal cost of the installation includes not only the array but a portion of an enlarged bottoming Rankine cycle generator, so that the cost will be very approximately equivalent to a low temperature, stand-alone solar plant of similar size with a steam turbine the direct costs may be lower if the overall CC plant is very large so that allocated power block costs are lowered to levels typical of large turbines. The LEC of such an arrangement will be clearly higher than that of an identical array used as a coal saver at the same site with no capital cost attribution, while the emissions replacement potential is less than for a coal saver. This is therefore not discussed further as a preferred option.

In Table 7.2.3 the cases of solar biomass boilers and wind are examined. For solar biomass boilers, the plant is assumed to replace coal-fired baseload plant. The direct generating cost in the example given is significantly higher than for coal fired generation, at US\$ 0.06 to US\$ 0.11 per kWh. Biomass capital costs are relatively high, with

Table 7.2.3:   Examples of new hybrid trough, CLFR and wind plants compared to a standard LS3 trough/gas hybrid

Plant type	LS3 Biomass Low	LS3 Biomass High	CLFR Biomass Low	CLFR Biomass High	Wind
Backup system	Biomass	Biomass	Biomass	Biomass	n/a
Biomass fuel cost	Low	High	Low	High	n/a
Location	Rockh.	Rockh	Rockh	Rockh	Rockh
Time frame/production scale	Mod. prod.	Mod. prod.	Mod. prod.	Mod. prod.	2000
Mirror aperture (m^2)	470880	470880	488100	488100	—
Circulating fluid temperature ($_i$C)	390	390	365	365	—
Net capacity (MW$_e$)	80	80	135	135	100
Ave global insolation (MJ/m^2/day)	20.0	20.0	20.0	20.0	—
Avg. daily solar output (MJ/m^2)	11.3	11.3	10.7	10.7	—
Turbine efficiency	0.37	0.37	0.37	0.37	—
Equivalent full load hours (h/yr)	7008	7008	7008	7008	—
Annual electric output (GWh/yr)	561	561	949	949	219
Solar share	0.36	0.36	0.21	0.21	.25 cap. fac.
Annual solar electric output (GWh/yr)	200	200	195	195	—
Direct costs (000s US$/yr)					
Site Works	included	included	included	included	included
Solar Field	included	included	included	included	included
HTF System/Boiler	included	included	included	included	included
Power Block (US$/kW$_e$)	2000	2000	2000	2000	included
Balance of Plant	included	included	included	included	included
Land	included	included	277	277	included
Indirect Costs	included	included	included	included	included
Project total (000s US$)	292346	292346	335564	335564	100000

(*Continued*)

Table 7.2.3:   (*Continued*)

	3654	3654	2477	2477	1000
Unit Cost US$/kW$_e$	3654	3654	2477	2477	1000
Biomass fuel cost (US$/GJ)	1.5	4	1.5	4	—
Fuel cost (US$/kWh)	0.016	0.043	0.016	0.043	—
Fuel cost (000s US$/yr, $\eta_{combustion} = 0.90$)	5849	15598	12225	32599	—
Annual O&M cost (000s US$)	4701	4701	5582	5582	1095
LEC (US$/MWh) — total	**79.4**	**104.6**	**62.6**	**93.7**	**52.9**
Capital cost fraction	52.1	52.1	35.4	35.4	45.7
Fuel cost fraction	15.1	40.3	18.7	49.8	0
O&M cost fraction	12.2	12.2	8.5	8.5	7.3
LEC US$/kWh — total	**0.079**	**0.105**	**0.063**	**0.094**	**0.053**
Yearly emissions avoidance cost					
Displaced fuel cost ($/kWh) — high	0.0300	0.0300	0.0300	0.0300	0.0300
Displaced fuel cost ($/kWh) — low	0.0025	0.0025	0.0025	0.0025	0.0025
Displaced capital cost ($/kW$_e$)	1140	1140	1140	1140	0
Displaced O&M cost ($/kWh)	0.0052	0.0052	0.0052	0.0052	0.0046
Total displaced cost ($/kWh) — high	0.067	0.067	0.067	0.067	0.050
Total displaced cost ($/kWh) — low	0.027	0.027	0.027	0.027	0.010
Net cost ($/kWh) — high	**0.012**	**0.037**	**−0.005**	**0.026**	**0.003**
Net cost ($/kWh) — low	**0.052**	**0.077**	**0.035**	**0.066**	**0.043**
Actual emissions (g CO$_2$/kWh)	50	150	50	150	0
Alternate emissions (g CO$_2$/kWh)	909	909	909	909	732
Emissions saved (t CO$_2$)	**546200**	**506100**	**916800**	**833100**	**184600**
Net cost/tonne CO$_2$ (high fuel cost)	**$12.4**	**$41.4**	**−$4.9**	**$30.0**	**$3.3**
Net cost/tonne CO$_2$ (low fuel cost)	**$53.4**	**$85.5**	**$36.4**	**$75.5**	**$50.6**

power block capital costs assumed to be \$2000 per kW$_e$. Biomass fuel costs can vary widely and so two extreme values of US\$ 1.50 and US\$ 4.00 per GJ are used as bounds. If the biomass plant operates from low-cost waste fuel, then solar will raise the cost of fuel, but this may still be beneficial because it can extend daily or seasonal operation and better utilize the capital facilities in the plant. Waste fuel is usually produced unsustainably, so the correct comparison for the long term would be to use sustainable plantation biomass costs. The lower fuel costs used here would be similar to plantation fuel costs in developing countries or in developed countries using high production forestry. The simple LECs calculated are higher than for coal savers in Table 7.2.2 but the net cost is comparable because the solar biomass is a high capacity factor plant offsetting coal-fired capacity.

Table 7.2.3 also shows an example of a wind plant installed in an area with a good resource with a capacity factor (CF) of 0.25 (most wind plant is installed with CF from 0.2 to 0.25). Installed capital cost of the plant is assumed to be 1000 US\$/kW$_e$, with O&M costs of 0.005 US\$/kWh [14]. Wind yields a lower LEC than the solar-biomass options but the net cost including displaced fuel costs is again comparable since there are no displaced capital costs. Both wind and solar options would have about 1/3 higher performance in optimal sites.

7.2.6   *Indirect costs and benefits of clean energy technology*

All technologies have external costs and benefits. It is economically efficient for these costs to be placed in a market form so that society can choose the least cost option. The present energy industry does not have to include its environmental and social costs in its pricing in most cases. The rest of society subsidizes these industries in health and environmental costs, and is less economically efficient as a result. What is required is not a specific subsidy of renewables, but proportional financial feedback to deter negative externalities.

In addition to direct costs, Tables 7.2.2 and 7.2.3 show the carbon emissions avoidance figures associated with each option. A better

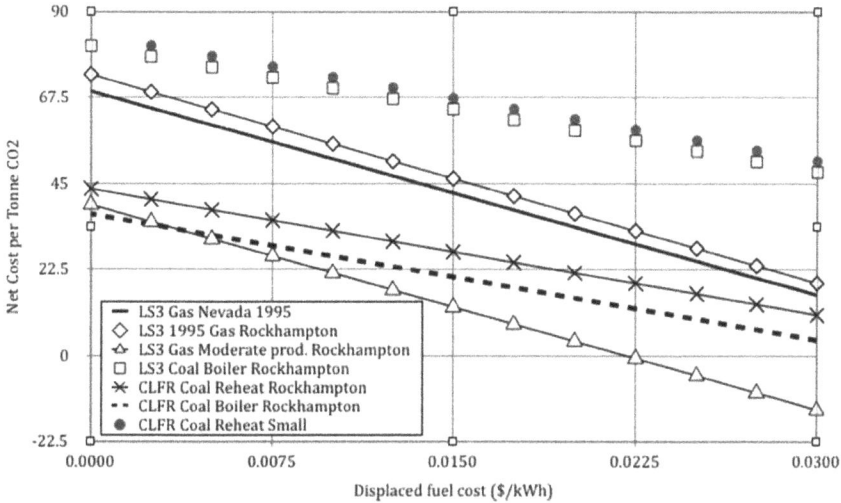

Figure 7.2.2: Estimated net cost per tonne of CO_2 replacement by LS3 and CLFR replacing coal fired electricity. The net cost drops to the right as the displaced coal fuel becomes more expensive. The CLFR cases shown do not have the advantage of gas backup to replace coal capital cost, but this could be provided. Negative cost per Tonne suggests a cost cheaper than coal generation

emissions avoidance figure would be achieved by offsetting brown coal-fired plant, but black coal plant is here used as the offset example for coal savers and hybrid solar plants which offset coal generation. The details behind these calculations are contained in Ref. [15]. CO_2 emissions reduction is expressed in US$ per Tonne in Fig. 7.2.2 for the cases in Table 7.2.2.

The results clearly show that retrofit coal savers are not only less expensive in direct cost terms, but environmentally even more cost-effective than solar/natural gas hybrids over a fuel cost range including most of the current coal market. An important point is that 100% of the solar array of a coal saver can be attributed to coal reduction because it is physically linked in the plant. The type of fuel that is replaced is an important issue that is frequently ignored.

The emissions avoidance costs for line focus coal savers shown are below those of the preferred tower + storage options described by Kolb [11] even though long term high production costs have been

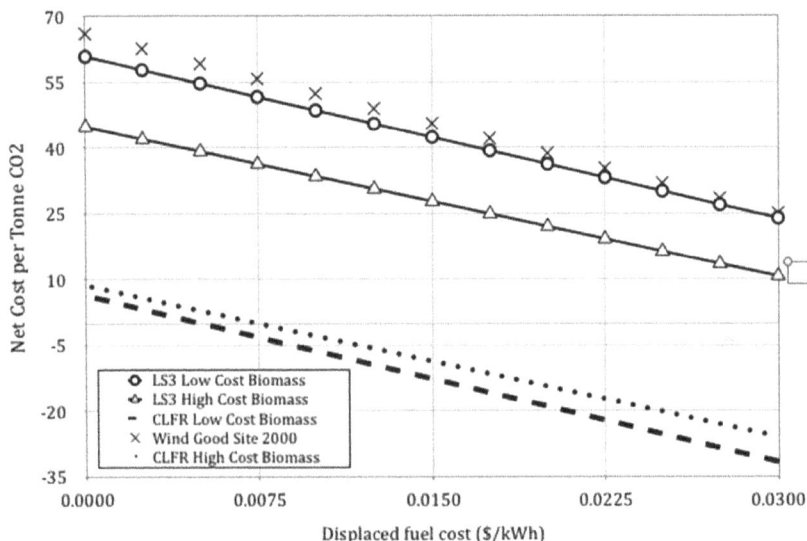

Figure 7.2.3:   Results of levelised CO_2 avoidance costs vs fossil fuel cost for solar-biomass line focus solar technologies and wind generation. High capacity factor (CF) solar-biomass options replace baseload coal, while low CF wind is assumed to replace 1/2 coal and 1/2 gas as used in a gas turbine. Biomass emissions will also drop as the transport economy becomes cleaner, and collection cost will drop as the industry develops in size. However, because wind offsets a grid mix of fuel, the fuel cost offset by wind may be higher than the cost of coal offset by biomass-solar plant

used for towers and not the line focus systems. This is done to emphasize that the line focus coal saver option is an immediate option for the industry, rather than a long term one. Coal saving addresses the immediate realities of a developed countries market belonging almost entirely to Market B (defined below in Section 7.2.7).

Figure 7.2.3 shows that the results for high capacity factor solar-biomass plants are also excellent and only slightly less favorable than for coal savers.

The results suggest that solar coal savers and solar/biomass plants will soon approach a figure of US$ 20 per tonne of avoided emissions, and because very high production runs are not presumed, there is further potential for reduction. This avoidance cost is likely to be higher than the avoided emissions cost using end use efficiency, but should be highly competitive against other generation technologies such as pure biomass and wind.

Global climate change is spurring development of financial methods of feedback (polluter pays) like carbon trading. In 1999, the World Bank published a consultant's report, referred to as *Enermodel* [16], on one particular technology, solar thermal electricity. The important result is that the inclusion of a carbon tax of only US$ 25 per tonne drops trough system STE by about 0.02 cents per kWh$_e$ to a figure similar to the cost of conventional baseload generation in the mid-transition market period (this is called long term by the Enermodel study authors, for whom 'long term' is 10–20 years). These figures will make solar-biomass and solar coal savers viable in many areas.

However, local pollutant effects could (and eventually should) also be included. Krewitt *et al.* [17] describe local environmental costs from coal-fired plants in the EU and Germany which range from 0.9 to 23 US cents per kWh$_e$, and produce an average estimate for 15 European countries of 6.4 cents per kWh$_e$. The results from Table 7.2.3 and the Eyre *et al.* [18] estimate suggest a combined global and local pollution cost of 5–180 US$ per tonne of CO_2. Stirling [19] summarizes a number of studies of local and global externality costs, most of which fall between 1 and 10 cents per kWh$_e$ in combination. Clearly, transition period STE is likely to be highly cost-effective if the most local pollution affected markets are addressed, and the full environmental costs of energy production in such markets are given market value.

7.2.7 *The current STE industry view*

Direct costs are the main factor determining whether STE generation technologies are economically viable. Kolb [11] has provided an excellent description of the long term potential of central tower options.

For parabolic troughs, DSG developments are projected to lead to a 5% reduction in total system cost [7]. However, this should be compared to the estimated cost reductions quoted in the same reference for improved plant system design (10%), larger plant size (10–15%), multiple plant procurement (10–15%), standardized engineering (5%) and competitive pressures (15–20%). Thus, the present day cost (~$US0.10 per kWh$_e$) of stand-alone solar trough/gas hybrids could be dropped to approximately half, but it is clear that increased market size and competition will have a far greater potential impact than solar plant technical improvements. In short, the establishment of large viable low cost transition markets with large scale plants will have a profound effect upon solar thermal electricity cost, just as it already has had upon wind generation cost.

Current hopes for increased production lie primarily with a few GEF based projects around the world set in developing countries. In the Parabolic Trough Roadmap [10] it is stated that *"near-term markets, although driven by a demand for solar power, will rely on GEF grants and other financial incentives to achieve cost parity with conventional power generation. Trough technology is likely to be integrated into larger combined-cycle plants to help improve the solar project economics."* However, in practice, GEF STE projects have proved to be very slow to come to fruition — none have yet done so. Moreover, as a market philosophy, the industry emphasis on GEF largely ignores the developed country markets in which the most massive reductions in emissions must occur.

An important issue is how STE industry growth can achieve highly sustained growth between now and 2035. The authors believe that the industry needs to widen its preferred market base enormously to achieve this. It needs to emulate the very large growth rate markets currently being addressed by wind in developed nations, and to develop the confidence that it can compete against wind and biomass in these markets.

Two broad types of markets exist for STE:

A-type markets: where there is a need for substantial new electricity capacity in the short term, as found in most developing countries;

B-type markets: where there is little or no short term need for new capacity, which is currently the case in Europe, North America and Australia.

It should be understood that there are some regions of developed countries where new capacity is required and some areas of developing countries where enough existing capacity exists, so a complete geographical identification is not possible. In the next two or three decades, most new plant will be in developing countries and most regions of oversupply in developed countries.

Let us examine one case. The integrated solar combined cycle system trough (ISCCS), referred to above, *en passant*, is often proposed for market A GEF projects. Is it the best option for Market B? Supplementation of combined cycle by trough plant is called 'power boosting' [11], and is characterized by a relatively constant fossil fuel output supplemented by a variable solar output making use of an oversized steam turbine. Kolb points out [11] that STE strategies which offset coal usage have the highest effectiveness in reducing emissions; but the option of supplementation by ISCCS non-storage trough plant delivers a fuel emissions offset typical of the grid mix of fuels, rather than exclusively coal. As a power booster, this option incurs an extra capital cost for enlargement of the capacity of the steam turbine. Retrofit fuel saver options which do not incur turbine capital costs and which offset pure coal should be superior to this if the same array type is used in both cases.

The current trough industry view is interesting for the medium and longer term. For the medium term the Parabolic Trough Roadmap [10] states *"Solar power opportunities will emerge where green markets materialize and mature. Mid-term markets will require the technology to achieve 6¢–8¢/kWh without special financial incentives other than a green electricity premium of 1¢–2¢/kWh_e. This technology will need to be dispatchable, preferably through thermal storage."*

Dispatchable? Green electricity — being exploited by wind — is largely a Market B phenomenon, for which storage is not needed. Green power schemes themselves usually contain no time of day requirement. Hence, it is important to emphasize non-storage hybrid options which may be sufficient to fuel perhaps the majority of early market growth.

For the long term, the Parabolic Trough Roadmap [10] states that *"Solar-trough power will need to become broadly competitive with conventional alternatives and will enjoy expanding markets globally throughout sunbelt regions. Long-term market opportunities will open up when the technology can compete at 4¢–5¢/kWh_e."* In the longer term, environmental and social costs are more likely to be included in both fossil fuel and renewable energy market cost, so that higher simple direct costs than this may be able to be tolerated in competition with fossil fuel, as will be discussed later in this presentation. On the other hand, other forms of renewable energy and end use energy efficiency share environmental advantages with STE. It is their long-term cost and market size compared to STE which represent the most important market bounds, not the cost of coal and oil.

7.2.8   New market niches

There are broad potential transition market niches available which may be more attractive than the ISCCS and solar/gas hybrids usually promoted by the industry.

7.2.8.1   Solar/biomass rankine cycle hybrids

These are envisaged as operating with a high solar fraction when solar is available, and using sustainable plantation biomass otherwise. Biomass would also be used for all superheating. Because biomass fuel can be scarce in some seasons, solar can be used to extend the annual capacity factor of biomass plant, and may reduce costs where plantation biomass is expensive. This option would be suitable for both developed and developing country use out to 2035, and it allows a 100% renewable energy fraction without use of thermal storage. Peak solar fractions up to about 90% are possible, but average solar fractions for 24 hour plant would be below 50%. Where coal plant is to be installed anyway (such as in India or China), large emissions reductions are possible through addition of a large solar array.

7.2.8.2 *Small solar fraction fuel saver retrofit to existing coal plant*

There is a very large existing global capacity of fossil fuel plant, much of it coal, which is available for low solar-fraction retrofit. STE systems can be used as simple coal savers [20, 21], which are solar-coal hybrids saving coal whenever solar energy is available. Coal fired plants tend to be much larger individually than solar thermal stand-alone plants (600–2,400 MW_e vs 5–80 MW_e) and useable land around coal fired plants can be restricted. Addition of steam below 285°C can replace bled steam from the turbines used in the reheat cycle, and solar steam at around 365°C can be introduced to the main boiler. Around the world, however, there are large numbers of such plants in good insolation areas, many with sufficient adjacent land area to accommodate a solar field of the size of the current largest STE units (80 MW_e), and all of them with turbo-generators installed and an existing attachment to the electricity grid.

Strikingly, in the two broad niches above, and for ISCCS evaporators and traditional solar/gas plant, simple low temperature, low cost linear absorber technology seems preferable and avoids the need for advanced technology storage systems and high temperature solar operation. This suggests that a very large initial focus should be made on such technology. However, this is not to say that high concentration collectors will not find their own market niches.

7.2.8.3 *High concentration niches*

Current research work into the development of tower receivers which can provide thermal energy for top cycle of a combined cycle plant [22], may be compatible with a trend to using combined cycle with biomass gasification instead of straight combustion. Also, the MTSA approach described later in this presentation moves concentrating solar to the end user market where its possibilities for imbedded high efficiency generation in the urban landscape allow the displacement of high retail costs and usage of waste heat to defray higher collector capital cost. Finally, the production of solar fuels will in some cases need high concentration. However, all of these are at an early stage of development in both collector field and system terms and will need to demonstrate their cost competitiveness against line focus options.

7.2.9   *Example of a coal saver — the CLFR*

In the design of solar thermal power plants there has been a tendency towards larger and larger scale systems to produce economies of scale and lower installation cost. However, with contiguous reflectors there are limits on the manageable size of such systems. Scaling up of parabolic trough or dish collectors for large solar thermal power systems is limited by wind loading problems and shading between adjacent concentrators. The aperture width of the Luz parabolic trough collectors [23, 24] is 5 m and the adjacent rows were spaced by approximately 10 m. Larger units become progressively more difficult to install and clean. Trough technology is now being developed by Pilkington [7], and work is proceeding on improvements to system components and maintenance in order to reduce system costs.

Another issue is that of heat transfer. One promising option for the reduction of the cost of current parabolic trough solar thermal power systems is to change to direct steam generation in the absorber. This removes the need for heat exchangers between the collector working fluid (usually oil) and the power cycle fluid (usually water). There has been considerable progress with the design of direct steam generation collectors. However, the implementation of this concept in concentrators such as the Luz trough collector is difficult because of the need for a flexible fluid coupling to the absorber and the requirement for turbulent flow in small diameter absorber pipes running for hundreds of meters. As the operating pressure of a direct steam generation collector will be in excess of 40 bar, the flexible hoses used in the existing hot oil working fluid collectors must be changed to rigid rotating couplings.

The basic concept of large reflectors being broken down into many Fresnel sub-elements to improve manageability was advanced by Baum *et al.* [25], and in the 1960s, important development work was undertaken by the important solar pioneer Giovanni Francia [26] of the University of Genoa. A Fresnel system can be designed with a stationary absorber so that a high-pressure direct steam generating absorber does not need flexible couplings. The system can also be scaled up in size without increasing the aperture width of the mirror components. A large Fresnel system can be designed with only one absorber line for a collector aperture width of 50 m or more, but as a Fresnel system is scaled up, the spacing between the outer lines of mirrors must be increased to avoid blocking between adjacent reflectors. This is a problem for trough and dish systems as well; the ground coverage used in the LUZ trough arrays is only 33%. The limited ground coverage of a classical Fresnel system can be overcome by a new configuration called the Compact Linear Fresnel Reflector (CLFR).

A simple linear Fresnel system has only one linear absorber on a single linear tower, and therefore there is no choice about the direction of orientation of a given reflector. But for technology supplying electricity in the multi-megawatt range, there will be many linear absorbers in the system. Thus, individual reflectors can have the option of directing reflected solar radiation to at least two absorbers in linear systems. This is the basis of the CLFR approach.

The additional variable in reflector orientation provides the means for much more densely packed arrays. This is not only because the boundary between continuous regions looking at one or the other reflector can be shifted and optimized as the sun moves during the day, but also because patterns of alternating row reflector inclination can be set up such that closely packed reflectors can be positioned with minimal shading and blocking, and also because domains associated with one or the other tower can enlarge or shrink with time to allow optimum collection. The interleaving of mirrors between two linear absorber lines is shown in Fig. 7.2.4.

This arrangement minimizes beam blocking between adjacent reflectors and allows higher reflector densities and lower absorber tower heights to be used. The advantages of interleaving become stronger as the reflector density in the field increases. However, even where reflector density is low, and interleaving is not required, the presence of variable size domains of reflectors allows closer packing than in a simple LFR.

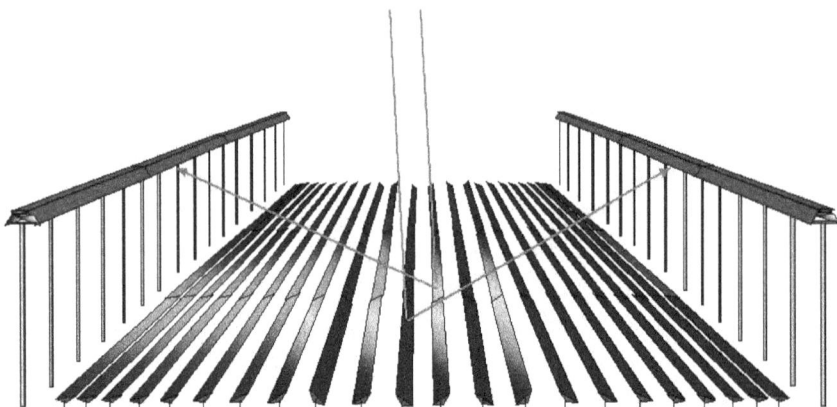

Figure 7.2.4:   Schematic diagram showing interleaving of mirror rows to achieve high site coverage without shading between adjacent mirrors. Mirrors in real array can be very close together

The CLFR power plant concept proposed for the Stanwell project is intended to reduce costs in all elements of the solar array. The following features enhance the cost effectiveness of this system compared with trough technology:

- Reflectors mounted close to the ground are used to minimize structural costs.
- Costly sagged glass reflectors are replaced by elastically curved glass reflectors.
- The absorber/heat transfer loop is isolated from the reflector field and does not move, thus avoiding the high cost of flexible high pressure lines or high pressure rotating joints required in trough and dish systems.
- Water/steam is used for heat transfer, and passive direct boiling heat transfer can be used to minimize parasitic pumping losses and the need for flow controllers. Steam supplied may either be directly into the power plant steam drum or via a heat exchanger. Steam can also be supplied in a similar manner for power plant reheating cycles.
- A simple cavity absorber design has been evolved and tested at a component test facility at Sydney University. There are also air

stable surfaces being developed at the University of Sydney and elsewhere that are suitable for higher temperatures. An inverted cavity absorber allows the use of large conventional steam pipes which can be attached to the plate. These are similar to the pipes used in conventional boilers and are relatively inexpensive.

- Low array maintenance costs due to ease of access for cleaning the ground mounted mirrors, and the capability to remove absorber panels without breaking into the heat transfer fluid circuit.

Stanwell Corporation, Solahart International, Solsearch Pty. Ltd. and the Universities of Sydney and New South Wales are working to commercialize solar energy by building a large CLFR solar thermal power generation plant at Stanwell Power Station near Rockhampton in Queensland. This plant, which will be the largest solar array in Australia, will produce a peak of 13 MW of thermal energy, which will offset the use of coal in the generation of electricity. The project is estimated to cost A\$7 million. Stanwell Corporation Limited (SCL), as manager for the project, has received a technology commercialization grant of A\$2 million from the Australian Greenhouse Office under the Greenhouse Showcase Program. The solar plant will be attached to a 1440 MW_e coal fired power station owned by SCL as shown in Fig. 7.2.5. More than 18,000 m^2 of mirrors will be built beside the power station.

The first CLFR plant will be used for heating feed water going into the reheating circuit, although subsequent plant will be able to be used for main boiler steam injection in to the cold reheat line. The design steam delivery conditions for the Stanwell project are 265°C and 5 MPa wet steam. Solar output will be able to be interrupted even if total plant output is constant.

The CLFR can, in principle, allow a peak output of 200MW_e per km^2 of ground area, but the current cost optimum is about 140 MW_e per km^2. As a comparison, an 80 MW_e LS3 plant in California occupies about 1.35 km^2, about 60 MW_e per km^2. Each of the four turbines can absorb about 30MW_e of additional heat input, so there is considerable room for expansion without consideration of main boiler steam supply.

Figure 7.2.5:　Artist's impression of CLFR plant at Stanwell

The first $4.2\,MW_e$ plant is being built for approximately \$US 1,000 per kW_e, a lower figure than other direct solar technologies of which the authors have knowledge. This is partially because only the array needs to be constructed and not electrical infrastructure. It is believed that this cost will drop substantially in future larger arrays.

7.2.10　*Beyond line focus — the MTSA*

The cost advantages of simple line focus collectors in primary generation niches described above is clear. However, in this section we examine a future collector possibility which may allow cost-effective use of highly concentrated solar energy.

Photovoltaic (PV) technology is the most well-known way to convert solar radiation into electric power even though it is not the most cost-effective way. In 1997 about 80% of direct solar electricity generated was supplied from concentrating solar Power (CSP) Plants using heat engine generators [27], almost all of it using line focus collectors.

CSP plants are mostly large power plants installed in sunny rural areas. However, conventional PV has the advantage that it can easily be installed in small units. This opens a market where one is able to invest money to install such units on his or her own roof. The Multi Tower Solar Array (MTSA) is a new kind of CSP plant which, like PV panels, can be integrated into the urban environment efficiently due to its overlapping heliostat fields [28]. The buyers can be either private persons who are aware of the environmental problems of conventional energy production, or institutions such as hospitals, factories or local governments. Since a small MTSA can be set up as a modular system, it has the additional advantages that it can be built step by step in stages or retrofitted to existing plants. This characteristic can reduce the risks in marketing solar power systems. An important area of application is the urban environment, where concentrating solar power plants can be installed on the roofs of big buildings such as industrial halls or shopping complexes or over open areas like parking lots. Unlike flat panel PV, this is not a technology for micro applications.

7.2.11 *Multi tower solar array technology*

The MTSA can be regarded as a group of solar towers where the heliostat fields of the towers partly overlap. It is in fact the two-axis tracking version of the single axis tracking CLFR, in which heliostats more distant from the towers are alternately directed to more than one tower. This leads to a high Annual Ground Area Efficiency if compared to a single tower system, where radiation would get lost because of the mutual blocking of the heliostats. Neighboring heliostats which are directed to two different towers have different blocking and shading properties [28].

The idea of the overlapping heliostat fields is, in principle, applicable to normal Solar Towers with heights of about 100 m. but this would lead to extremely large arrays. The idea in this paper

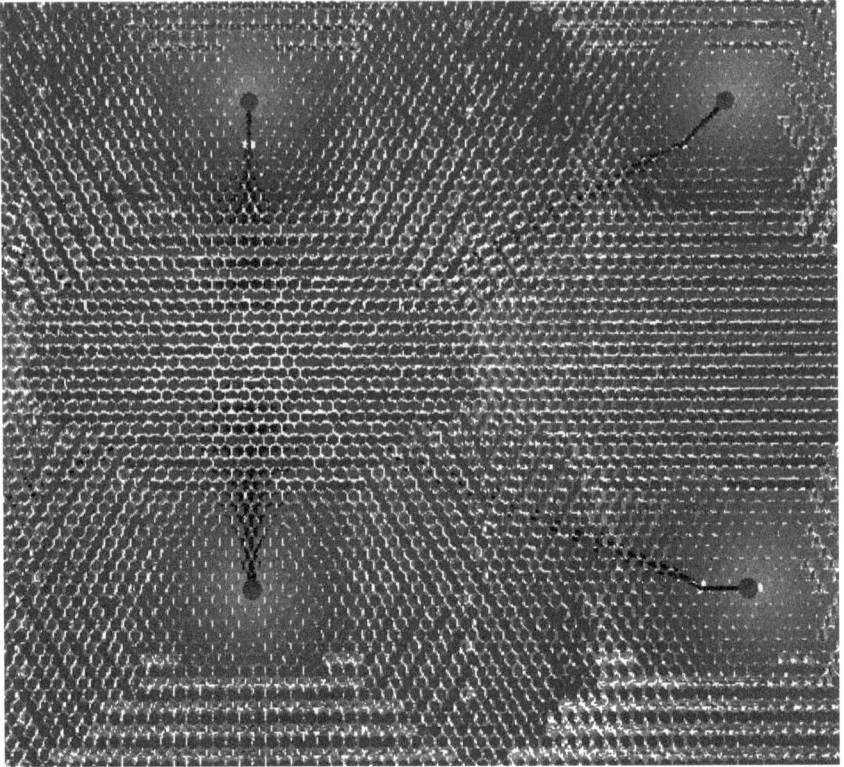

Figure 7.2.6:　Ray trace of portion of MTSA array as seen from above

is to design a system with several small towers instead of one high tower. The height of the towers can be designed depending on the given conditions. For applications in an urban environment, the small towers would have sizes between 5 and 10 m, comparable to large lampposts. Figure 7.2.6 shows the appearance of an MTSA with hexagonal reflectors.

There can be many different ways to use the concentrated radiation with an MTSA. Beside the production of hot steam for a turbine, it would be possible to use PV receivers, or heat engines such as Stirling engines to generate electricity from the concentrated radiation. Further applications could include the production of process heat or the use of the highly concentrated radiation for solar chemistry applications such as solar detoxification, solar reforming of methane, and production of hydrogen [29, 30, 31].

More recently, it has become evident that the MTSA may be an excellent vehicle for paired receivers of different type. In this concept, solar beam radiation from the primary reflectors strikes a special panel called a beam splitter or band-pass filter before striking the absorber. The beam is split spectrally into two beams, one of which strikes a PV receiver and the other a receiver which can make use of thermal energy.

Why do this? The PV receiver operates by having solar photons boost electrons in a semiconductor into the conduction band. Parts of the solar spectrum in which photons do not have enough energy to boost the electrons in this way are simply wasted, but in parts of the spectrum toward the blue wavelengths, the photons will also have far too much energy and after the electron has been promoted, the excess will be wasted and will only serve to heat up the absorber. The special filter proposed would direct to the PV panel only the wavelengths which can be used efficiently, and direct the remainder to either a heat engine or methane reformer, which can use any wavelength. PV receivers which have very high efficiency are now becoming available for concentrating systems, but even a restricted wavelength silicon PV receiver can achieve efficiencies close to 40%.

The ultimate aim is to use the collected solar beam to the best advantage, and this principle can be carried further. Because the MTSA is near the site of the end use of energy, waste heat from the heat engine and heat from the PV absorber can be used to supply energy for air conditioning, building heating, or water heating. In this way up to 80% of the available energy in the beam can be utilized. This is much higher than in existing trough or CLFR technology, which is unlikely to exceed peaks of 40% because of limitations in heat engine efficiency and difficulty in using waste heat in remote generation locations.

A primary or secondary focus also has the potential to run a methane to hydrogen reformer with the collected thermal energy.

This could provide hydrogen fuel for cars or provide stored energy for electricity. The latter option would convert the direct solar MTSA into a high capacity factor plant.

Greatly increased efficiency in the use of beam radiation comes at a price, because an MTSA will have increased field costs, and costs for special equipment like advanced PV, a beam splitter, two axis reflector tracking as an addition. However, initial estimates of cost performance suggest that the economic viability is similar to low cost rural generating plant like the CLFR 3. An important benefit of the urban MTSA is that it can offset the retail cost of fuel, rather than the generating cost. There is often a factor of two advantage over generation cost. But even the latter is expected to be much less than that for electricity produced with conventional PV technology.

Aesthetically, an MTSA would feature towers comparable in size and appearance to large lamp posts which can be well integrated in an urban environment due to their small scale. Seen from above, the reflectors would provide a changing image of the sky, but not the glare of the solar disk. In more equatorial countries, parking sites do have the problem that the parked cars overheat in direct sun. An MTSA reflector field over such a parking site could utilize the solar radiation and simultaneously protect the cars from overheating.

From the architectural point of view, the MTSA would work like a day-lighting system which allows diffused daylight to illuminate the interior of the building by protecting it from the direct solar radiation. In that way it can be compared to a saw-tooth roof facing away from the equator, which allows only diffused daylight into the building. An MTSA could be used over large buildings or open areas like parking sites. It could provide protection from the sun and could utilize the solar radiation to produce solar electricity and heat at the same time.

The MTSA offers the new possibility of Concentrating Solar Power Plants being set up in an urban environment. Using small scale towers, such a system can be well integrated into the architectural context of a city. From the street, such systems could be installed in such a way on roofs that they would be barely recognized from ground level. On the other hand, the towers and the field of bright — but not glaring — reflectors offer new architectural possibilities where the reflector field is visible. The motion of the sun would be clearly evident in the constantly changing patterns of the heliostats and their sky images during the day. This would be a "living" roof, slowly changing with the time of day and the seasons.

7.2.12 *Summary*

In a recently submitted paper [2] we suggest that direct solar electricity is unlikely to contribute significantly to global emissions reduction before 2035, but becomes essential to meeting climate goals after this time. To fulfil this long term role, rapid growth must ensue now and be continued for decades so that full market share can be achieved. With the above background, in this chapter we argue for a new industry market approach which has important differences from current STE industry visions for the future.

The STE industry advocates installation of ISCCS and solar/gas hybrids as Global Environmental Facility (GEF) projects in developing countries. These are not the most emissions-effective modes of installation, nor do they address the very large emissions 'debt' in developed nations.

The main features of the suggested industry approach outlined in the chapter can be summarized as follows:

1. The primary market for STE should mostly be in developed countries rather than in developing countries under GEF projects.
2. Until 2035, transition markets which use hybridization but not thermal or chemical storage are recommended to create strong STE industry growth. These applications can all use low cost line focus technology. They are:

   - Solar biomass hybrids with a significant solar fraction to provide firm capacity in both developed and developing countries.
   - Solar fossil (mainly coal) hybrids with a significant solar fraction for use in developing countries.

- Solar coal savers supplying steam for fossil fueled plants in developed nations at a low solar fraction.

In addition, future possibilities like the MTSA may offer comparable cost effectiveness in new niches such as integrated electricity and heat or electricity and hydrogen fuel supply at the end use site in urban areas.

Evidence presented in this paper suggests that well targeted solar thermal electricity and other renewable energy options are likely to be less expensive as a total societal cost than conventional fuel, and may be highly competitive against pure wind and biomass technology without the invocation of storage. The main bounds on future growth will come from the performance of renewable energy competitors rather than fossil fuel. Future storage options applied to line focus technology may improve CO_2 avoided cost, as will large scale production.

This presentation does not address policy issues, but it is essential to the development of the STE industry that full life cycle costing information be adequately developed as an essential energy policy input so that equitable societal costing of different energy options can be performed. Direct solar energy has immense potential as a major energy source if we work to fully exploit potential market niches using new and exciting technology.

A fair and full costing of all energy is the vital ingredient to its success. The present system, in which hidden subsidies to fossil fuel are endemic both in terms of infrastructure support and a community write-off of environmental damage, must not be accepted as given. This practice grossly distorts both renewable energy technology development and public views of the effectiveness of renewable energy.

7.2.13 *Acknowledgments*

The authors wish to acknowledge the contribution of the financial assistance of His Royal Highness Prince Nawaf Bin Abdul Aziz of the Kingdom of Saudi Arabia through the Science Foundation for Physics within the University of Sydney.

7.2.14 *References*

[1] *Intergovernmental Panel on Climate Change*, 1995, IPCC Second Assessment: Climate Change, IPCC, Geneva, p. 19.

[2] Mills D.R. and Dey C.J., 2000. Renewable energy technology mix and atmospheric carbon dioxide stabilisation by 2050. Advances in Solar Energy **14** (In press).

[3] IEO(1999). *International Energy Outlook*, 1999. Energy Information Administration, Office of Integrated Analysis and Forecasting, U.S. Department of Energy, Washington, DC 20585, March 1999 (also available at http://www.eia.doe.gov/emeu/international/contents.html).

[4] Mills DR and Dey, C.J., *Solar Energy — the State of the Art*, 2001. Chapter 11 on Solar Thermal Power. Ed. J. Gordon. ISBN 1 902916 23 9. James and James, 35-37 William Rd. London.

[5] Alsema E.A., Frankl P., and Kato K., 1998, Energy Payback Time of Photovoltaic Energy Systems: Present Status and Prospects, in: *Proc. 2nd World Conference and Exhibition on PV Solar Energy Conversion* (J. Schmid *et al.*, eds.). Report EUR 18656 EN, 2125-2130.

[6] Thomas M.G., Post H.N. and DeBlasio R., 1999. Photovoltaic systems: An end-of-millennium review, *Progress in Photovoltaics: Research and Applications* **7**, pp. 1–19.

[7] Pilkington, 1996. Status report on solar thermal power plants. Pilkington Solar International GmbH, Germany.

[8] Cohen G.E, Kearney D.W. and Price H.W., 1999. Performance history and future costs of parabolic trough solar electric systems. In Proceedings of the 9th SolarPACES International Symposium on solar Thermal Concentrating Technologies, 22–26 June 1998, Font-Romeu, France, pp. 169–179, *Journal de Physique* IV, **9**, EDP Sciences, Les Ulies cedex A.

[9] Price H. and Kearney D., 1999. Parabolic-trough technology roadmap: A pathway for sustained commercial development and deployment of parabolic-trough technology. Sunlab, January 1999. Available at the Sunlab Website http://www.eren.doe.gov/sunlab/Frame_4.htm.

[10] *Parabolic Trough Roadmap*, 1999. Parabolic-trough technology roadmap: A pathway for sustained commercial development and deployment of parabolic-trough technology. Sunlab, January 1999. Available at the Sunlab URL http://www.eren.doe.gov/ sunlab/Frame_4.htm.

[11] Kolb G.J., 1998. Economic evaluation of solar-only and hybrid power towers using molten-salt technology, *Solar Energy* **62** (1), 51–61.

[12] Pacheco J.E., Reilly H.E., Kolb G.J. and Tyner C.E., 2000. Summary of the Solar Two test and evaluation program. *Proceedings of the 10 SolarPACES International Symposium on Solar Thermal Concentrating Technologies, Sydney, Australia, 8–10 March 2000*, pp. 1–11.

[13] Mills D.R. and Morrison G.L., 2000. Compact linear Fresnel reflector solar thermal power plants, *Solar Energy* **68** (3), 263–284.

[14] BTM Consult, 1999. International Wind Energy Development; World Market Update 1998. Published by BTM Consult ApS, Denmark.

[15] Mills D.R. and Dey C.J., 2000b. Development Strategies for Solar Thermal Electricity Generation. *Advances in Solar Energy* **14** (In press).

[16] *Enermodal*, 1999. Cost Reduction Study for Solar Thermal Power Plants, prepared for The World Bank by Enermodal Engineering Limited 650 Riverbend Drive, Kitchener, ON N2K 3S2, Ontario Canada, May. Available for download at http://www.eren.doe.gov/sunlab/Frame_4.htm.

[17] Krewitt W., Heck T., Trukenmuller A. and Rainer F., 1999. Environmental damage costs from fossil electricity generation in Germany and Europe, *Energy Policy* **27** (3), 173–183.

[18] Eyre N., Downing T., Hoekstra R., Rennings K. and Tol R., 1997. Global warming damages ExternE global warming subtask. Final Report. European Commission DG XII, JOULE Contract JOS3-CT95-0002. Referenced in Krewitt *et al.* (1999).

[19] Stirling A., 1997. Limits to the value of external costs, *Energy Policy* **25** (5), 517–540.

[20] Zoschak R.J. and Wu S.F., 1975. Studies of the direct input of solar energy to a fossil-fueled central station steam power plant, *Solar Energy* **17** (5), 297–305.

[21] You, Y. and Hu, E., 1998. Thermodynamic advantages of using solar energy in a regenerative Rankine cycle power plant, *Applied Thermal Engineering* **19**, 1173–1180.

[22] Kribus, A. *et al.*, 1999. A multistage solar receiver; the route to high temperature. *Solar Energy* **67** (1–3) 3–11.

[23] Kearney D., Kroizer I., Miller C. and Steele D., 1985. Design and Preliminary Performance of Solar Electric Generation Station I at Daggett-California, *Proc. 9th Biennial ISES Congress*, pp. 1424-1428, Canada.

[24] Jaffe D., Friedlander S. and Kearney D., 1987. The LUZ Solar Electric Generating System in California., *Proc. ISES Solar World Congress*, pp. 519–529, Germany.

[25] Baum V. A., Aparasi R. R. and Garf B.A., 1957. High Power Solar Installations, *Solar Energy* **1**, 6–13.

[26] Francia, G., (1968. Pilot plants of solar steam generation systems, *Solar Energy* **12**, 51–64.

[27] Trieb F., Langniß O. and Klaiß H., 1997. Solar Electricity Generation — a comparative view of Technologies, Costs and Environmental Impact, *Solar Energy* **59**, 89–99.

[28] Mills D.R. and Schramek P., 1999. Multi Tower Solar Array (MTSA) with ganged heliostats. In Proceedings of the 9th SolarPACES International Symposium on solar Thermal Concentrating Technologies, 22–26 June 1998, Font-Romeu, France, pp. 83–88, *Journal de Physique IV*, **9**, EDP Sciences, Les Ulies cedex A.

[29] Bett A.W., Stollwerck G., Sulima O.V. and Wettling W., 1998. Highest Efficiency GaAs/GaSb Tandem Concentrator Module. In Proceedings of 2nd World Conference on Photovotaic Solar Energy Conversion, July 1998, Vienna.

[30] Winter C.-J., Sizmann R.L. and Vant-Hull L.L., 1991. *Solar Power Plants* Springer Verlag Berlin Heidelberg.

[31] Flamant G., Ferrière A. and Pharabod F., 1999. *Proceedings of the 9th SolarPACES International Symposium on solar Thermal Concentrating Technologies*, 22–26 June 1998, Font-Romeu, France, pp. 247–383, *Journal de Physique IV*, **9**, EDP Sciences, Les Ulies cedex A.

7.3   Discussion Following David Mills's Presentation

Gordon: From 13 MW thermal input your system might give 6 MW of cooling power. However, if you were to use the same thermal input to drive an off-the-shelf chiller with COP = 1.4, then you could get close to 20 MW of cooling power, the typical cooling requirements for a suburban shopping mall. I am surprised that there is a difference here, not of a few percent, as would be the case if one compares one possible strategy with another, but of a factor or 3-4.

Mills: The difference must, of course be of only a few percent when one compares two reasonable alternative strategies. The first path to cooling might be to use the electrical output (4.2 MW) to run a vapor compression cycle of COP = 4, yielding 16 MW of cooling power. If it were a reasonably up-to-date vapor compression unit, this could rise to about 20 MW. The second, as you suggest, is to use a double effect absorption system at COP = 1.4, yielding 18 MW of cooling. So the two methods yield the same approximate output. Absorption may gain a slight practical advantage in allowing lower collector temperatures and pressures.

The more important difference between the systems is an assessment of the relative capital and running costs of a heat engine/generator/vapor compression cycle vs. the cost of an absorption cycle. Both systems are capable of storing cooled fluid for later use.

One more point: with new Brayton microturbines the waste heat output is near 250°C and an absorption cycle will be preferred in this case because the heat engine efficiency will be low. You could use both systems for microturbines, vapor compression using the electrical output and absorption using the waste heat. This may lead to a net system cooling efficiency higher than the above.

Roy: Your claim of $1/W sounds too good to be true. Would you please elaborate on how this can be achieved.

Mills: This is the cost of performing a solar retrofit to an existing system. It does not include the cost of a turbine or any other components that are assumed to be present. Nevertheless, I agree that it is still remarkably low. We achieve such a low cost due to the simplicity of our design and economies of scale. For example, we use extremely simple absorber technology which includes normal boiler pipes. Also, our direct steam generation is simple in the extreme: we don't care whether the pipe insides are partially or fully wetted or whether there is slug flow. Structural costs are low because our structures are close to the ground. Also, a surprising result of our study was that it is cheaper to employ labor than machines for keeping the collectors clean.

Roy: What is the initial cost of the reflectors?

Mills: It works out at approximately US$ 76 per m^2.

Ries: That is remarkably low. How are the mirrors made?

Mills: From extremely thin glass sheets, manufactured in Switzerland and silver coated in Australia. Because of the extreme thinness, one does not require heat for shaping the glass. Simple pressing and gluing will suffice.

Ries: That may be true for 2-dimensional surfaces but surely not for 3-dimensional curving, unless the pieces are extremely small.

Mills: The cost may be higher for 3D surfaces but the technique is the same. And the size of the individual pieces that may be used is really only limited, at present, by the dimensions the glass manufacturer can supply (1400 mm before cutting).

Chapter 8

2002

8.1 Editor's Foreword

When we think of solar power production, there is a tendency to think only of electricity generation, conveniently forgetting that this form of energy constitutes only approximately one-third of the world's consumption of fossil fuel. For this reason, from time to time, the Sede Boqer Symposia included key-note lectures on a number of related — either directly, or indirectly — subjects. One important potential large-scale use of solar-thermal energy, as illustrated in Aldo Steinfeld's keynote lecture, might be for the production of transportation fuel, which constitutes roughly 25% of global fossil fuel usage.

8.2 Solar Thermochemical Production of Solar Fuels (Prof. Aldo Steinfeld)

Invited keynote presentation given by Professor Aldo Steinfeld, Institute of Energy Technology, ETH, CH-8092 Zurich, Switzerland, and Paul Scherrer Institute, CH-5232, Villigen, Switzerland.

8.2.1 *Abstract*

This presentation develops some of the underlying science and describes some of the latest technological developments for converting solar energy into storable and transportable chemical fuels. The thermodynamics of solar thermochemical conversion and the most promising solar thermochemical processes are discussed. The reader is referred to [1] for a more comprehensive treatment of the solar thermochemical process technology.

8.2.2 Nomenclature

C	solar flux concentration ratio
I	normal beam insolation
HHV	high heating value
Q_{solar}	total solar power coming from the concentrator
T	nominal solar cavity-receiver temperature
ΔG	Gibbs free energy change per mole of reactant
ΔH	enthalpy change per mole of reactant
ΔS	entropy change per mole of reactant
$\eta_{absorption}$	solar energy absorption efficiency
η_{Carnot}	efficiency of a Carnot heat engine operating between T_H and T_L
η_{exergy}	exergy efficiency
σ	Stefan–Boltzmann constant (5.6705×10^{-8} Wm^{-2}K^{-4})

8.2.3 *Thermodynamics of solar thermal conversion*

Figure 8.2.1 illustrates the basic idea of solar thermochemical conversion. If we concentrate the diluted sunlight over a small area with the help of parabolic mirrors and then capture that radiative energy with the help of suitable receivers, we would be able to obtain heat at high temperatures for driving a endothermic chemical transformation. The solar energy absorption efficiency of a solar reactor, $\eta_{absorption}$ is defined as the net rate at which energy is being absorbed divided by the solar power coming from the concentrator. Solar reactors for highly concentrated solar systems usually feature the use of a cavity-receiver type configuration, i.e. a well-insulated enclosure with a small opening (the *aperture*) to let in concentrated solar radiation. At temperatures above about 1,000 K, the net power absorbed is diminished mostly by radiative losses through the aperture. For a perfectly insulated blackbody cavity-receiver, it is given by [2]

$$\eta_{absorption} = 1 - \left(\frac{\sigma T^4}{IC} \right). \tag{8.1}$$

where C is the mean solar flux concentration, I is the normal beam insolation, T is the nominal cavity-receiver temperature, and

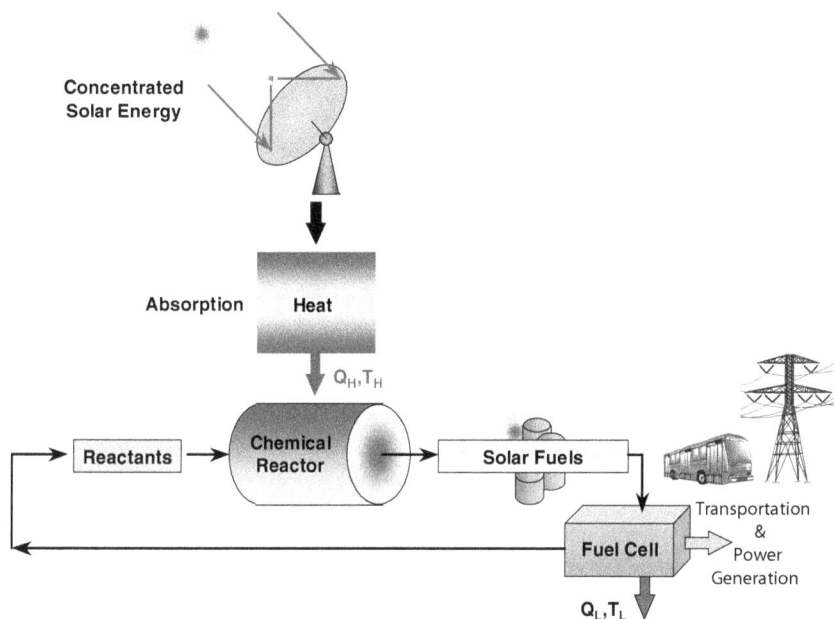

Figure 8.2.1: Schematic of solar energy conversion into solar fuels. Concentrated solar radiation is used as the energy source for high-temperature process heat to drive chemical reactions towards the production of storable and transportable fuels

σ the Stefan–Boltzmann constant. The absorbed concentrated solar radiation drives an endothermic chemical reaction. The measure of how well solar energy was converted into chemical energy for a given process is the exergy efficiency, defined as

$$\eta_{exergy} = \frac{-\dot{n} \, \Delta G_{rxn}|_{298\,K}}{Q_{solar}}, \tag{8.2}$$

where Q_{solar} is the solar power input and ΔG_{rxn} is the maximum possible amount of work that may be extracted from the products as they are transformed back to reactants at 298 K. The Second Law is now applied to calculate the maximum exergy efficiency $\eta_{exergy,ideal}$. Since the conversion of solar process heat to ΔG_{rxn} is limited by both the solar absorption and Carnot efficiencies, the maximum exergy

η **exergy,ideal**

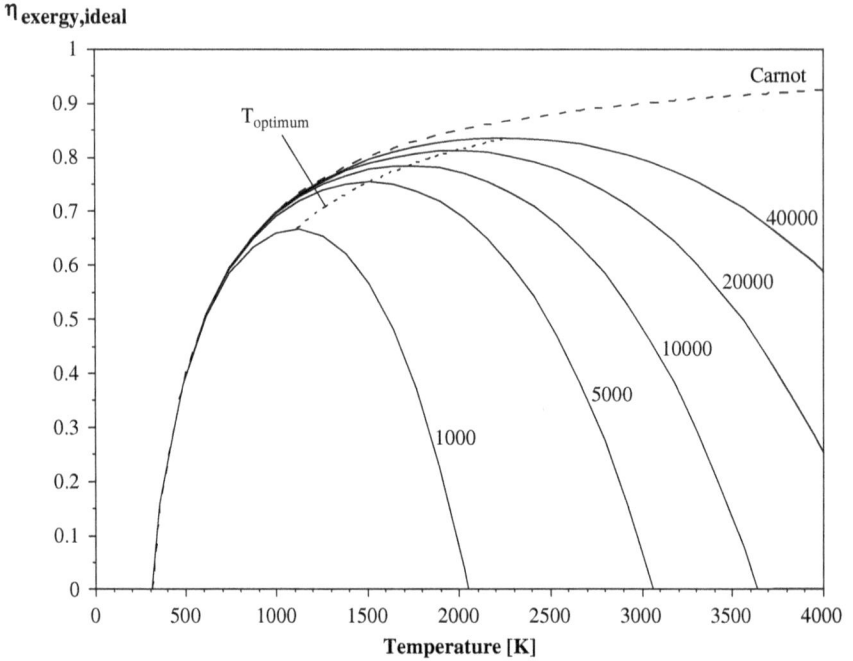

Figure 8.2.2: Variation of the ideal exergy efficiency $\eta_{exergy,ideal}$ as a function of the operating temperature T_H, for a blackbody cavity-receiver converting concentrated solar energy into chemical energy. The mean solar flux concentration is the parameter: 1,000, 5,000 ... 40,000

efficiency is

$$\eta_{exergy,ideal} = \eta_{absorption} \cdot \eta_{Carnot} = \left[1 - \left(\frac{\sigma T_H^4}{IC}\right)\right] \times \left[1 - \left(\frac{T_L}{T_H}\right)\right]$$

$$(8.3)$$

where T_H and T_L are the upper and lower operating temperatures of the equivalent Carnot heat engine. $\eta_{exergy,ideal}$ is plotted in Fig. 8.2.2 as a function of T_H for $T_L = 298\,\text{K}$ and for various solar flux concentrations. There is an optimum temperature $T_{optimum}$ for maximum efficiency, which varies between 1,100 and 1,800 K for uniform power-flux distributions with concentrations between 1,000 and 13,000 [3].

8.2.4   *Solar hydrogen production*

Single-step thermal dissociation of water, also known as water thermolysis, although conceptually simple, has been impeded by the need of a high-temperature heat source at above 2,500 K for achieving a reasonable degree of dissociation, and by the need of an effective technique for separating H_2 and O_2 to avoid recombination or end up with an explosive mixture. Among the ideas proposed for separating H_2 from the products are effusion separation [3, 4], and electrolytic separation [5, 6]. Water-splitting thermochemical cycles bypass the H_2/O_2 separation problem and further allow operating at relatively moderate upper temperatures. Previous studies performed on H_2O-splitting thermochemical cycles were mostly characterized by the use of process heat at below about 1,300 K, available from nuclear and other thermal sources. These cycles required multiple steps (more than two) and suffered from inherent inefficiencies associated with heat transfer and product separation at each step. A status review on multi-step cycles is given in [7].

In recent years, significant progress has been accomplished in the development of optical systems for large-scale collection and concentration of solar energy, using solar tower and tower-reflector technology at power levels of several megawatts and capable of achieving power flux intensities equivalent to solar concentration ratios of 5,000 suns and higher. Such high radiation fluxes allow the conversion of concentrated solar radiation to thermal reservoirs at 2,000 K and above. Thus, the door was opened for the more efficient two-step thermochemical cycles for splitting water using solar energy.

8.2.4.1   *Thermochemical cycles based on metal oxide redox reactions*

Several two-step water-splitting cycles based on metal oxide redox reactions have been proposed [8, and literature cited therein]. The first, endothermic, step is the solar thermal dissociation of metal oxide to the metal or the lower-valence metal oxide. The second, non-solar, exothermic, step is the hydrolysis of the metal at moderate

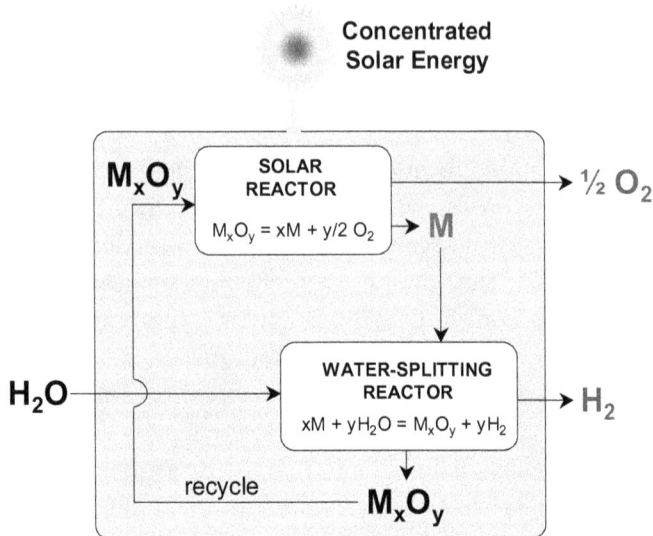

Figure 8.2.3:   Schematic of a two-step water-splitting thermochemical cycle using metal oxides in redox systems. In the first, endothermic, solar step, the metal oxide M_xO_y is thermally decomposed into the metal M and oxygen. Concentrated solar radiation is the energy source for the required high-temperature process heat. In the second, exothermic, non-solar step, the metal M reacts with water to produce hydrogen. The resulting metal oxide is then recycled back to the first step

temperatures to form hydrogen and the corresponding metal oxide. The net reaction is $H_2O = H_2 + 0.5O_2$. Hydrogen and oxygen are formed in different steps, thereby eliminating the need for high-temperature gas separation. This 2-step solar thermochemical cycle is shown schematically in Fig. 8.2.3. These cycles have been thermodynamically examined and tested in solar reactors for ZnO/Zn and Fe_3O_4/FeO redox pairs [8–13].

Other redox pairs, such as TiO_2/TiO_x, Mn_3O_4/MnO, and Co_3O_4/CoO have also been considered, but the yield of H_2 has been too low to be of any practical interest. Partial substitution of iron in Fe_3O_4 by other metals forms mixed metal oxides of the type $(Fe_{1-x}M_x)_3O_4$ that may be reducible at lower temperatures than those required for the reduction of Fe_3O_4, while the reduced phase $(Fe_{1-x}M_x)_{1-y}O$ remains capable of splitting water [14, 15].

Of special interest is the solar thermochemical cycle based on ZnO/Zn redox reactions, represented by:

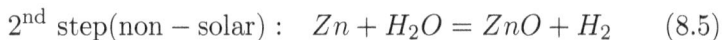

$$1^{st} \text{ step(solar)}: \qquad ZnO = Zn + 0.5O_2 \qquad (8.4)$$

$$2^{nd} \text{ step(non} - \text{solar)}: \quad Zn + H_2O = ZnO + H_2 \qquad (8.5)$$

Several chemical aspects of reaction (8.4) have already been investigated [11, 16–20]. At 2,235 K, $\Delta G° = 0$. The product gases needed to be either quenched or separated at high temperatures to prevent their recombination. Based on these previous studies and on the constraints imposed by the chemistry of the decomposition reaction, a solar chemical reactor was designed and a 10 kW prototype was fabricated and recently tested [21]. Its schematic configuration is shown in Fig. 8.2.4. It features a windowed rotating cavity-receiver lined with ZnO particles that are held by centrifugal force. With this arrangement, ZnO is directly exposed to high-flux solar irradiation and serves simultaneously the functions of radiant absorber, thermal

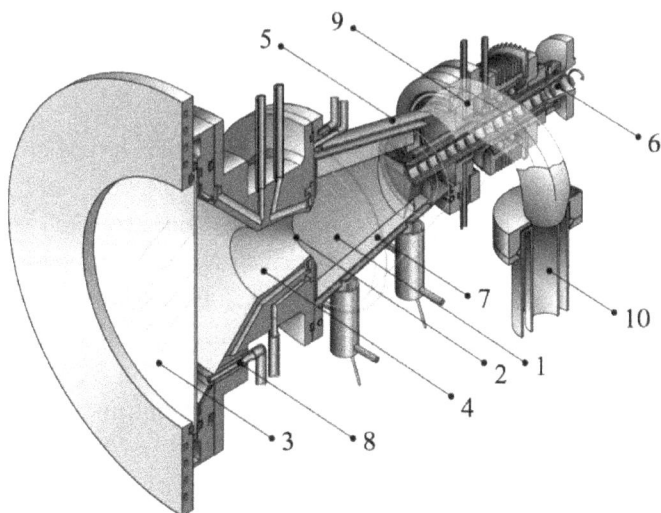

Figure 8.2.4: Schematic of a solar chemical reactor for the thermal decomposition of ZnO at 2,000 K. 1 = rotating cavity-receiver, 2 = aperture, 3 = quartz window, 4 = CPC, 5 = outside conical shell, 6 = reactant feeder, 7 = ZnO layer, 8 = purge-gas inlet, 9 = product outlet port, 10 = quench device [21]

insulator, and chemical reactant. The reactor development work is in progress and currently focused on experimental evaluation.

As far as the zinc hydrolysis is concerned, reaction (8.5), laboratory studies on the kinetics and preliminary tests with a novel hydrolyser concept indicate that the water-splitting reaction proceeds exothermally at reasonable rates when steam is bubbled through molten zinc at above about 700 K [22]. The technical feasibility of the solar reactor and hydrolyser at an industrial scale needs demonstration. An economic assessment indicates that the solar ZnO/Zn water-splitting cycle can become competitive with the electrolysis of water using solar-generated electricity [20].

8.2.4.2   *Carbothermic reduction*

The use of reducing agents brings about reduction of the metal oxides at much lower temperatures. The carbothermal reduction of metal oxides, using C(gr) and CH_4 as reducing agents, may be represented as:

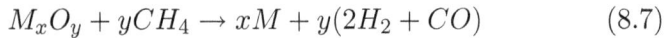

$$M_xO_y + yC(gr) \rightarrow xM + yCO \qquad (8.6)$$

$$M_xO_y + yCH_4 \rightarrow xM + y(2H_2 + CO) \qquad (8.7)$$

Examples of carbothermic reduction processes that have been carried out in solar furnaces include the production of Fe, Mg, and Zn from their metal oxides in Ar atmospheres, the production of AlN, TiN, Si_3N_4, and ZrN from their metal oxides in N_2 atmospheres, and the production of Al_4C_3, TiC, SiC, and CaC_2 from their metal oxides in Ar atmospheres [1, 23, 24]. Using natural gas as a reducing agent combines in a single process the reduction of metal oxides with the reforming of methane for the co-production of metals and synthesis gas (syngas), [see Eq. (8.7)]. Thermal reductions of Fe_3O_4 and ZnO with CH_4 to produce Fe, Zn, and syngas have been demonstrated in solar furnaces using fluidized bed and vortex type reactors [25–28].

8.2.4.3   *Solar decarbonization and upgrade of fossil fuels*

The substitution of fossil fuels with solar fuels is a long-term goal requiring the development of novel technologies. Strategically, it is desirable to consider mid-term goals aiming at the development of hybrid solar/fossil-fuel endothermic processes in which fossil fuels are used exclusively as chemical reactants and solar energy as the source of process heat. The products of such hybrid processes are cleaner fuels whose quality has been solar-upgraded: their calorific value is increased by the solar input in an amount equal to the enthalpy change of the reaction.

The mix of fossil fuels and solar energy creates a link between today's fossil-fuel-based technology and tomorrow's solar chemical technology. It also builds bridges between present and future energy economies because of the potential of solar energy to become a viable economic path once the cost of energy will account for the environmental externalities from burning fossil fuels. The transition from fossil fuels to solar fuels can occur smoothly, and the lead-time for transferring important solar technology to industry can be reduced.

An important category of thermochemical processes for mixing fossil fuels and solar energy is the decarbonization of fossil fuels, i.e. the removal of carbon from fossil fuels prior to their use for power generation. Two methods are considered: a) the solar thermal decomposition, and b) the steam-reforming/gasification. Both methods make use of high-temperature solar process heat for driving the endothermic transformations and are schematically shown in Fig. 8.2.5 in the form of simplified process flow diagrams.

The thermal decomposition of natural gas, oil, coal, and other hydrocarbons can be represented by the simplified reaction:

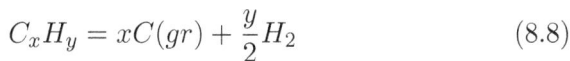

$$C_xH_y = xC(gr) + \frac{y}{2}H_2 \qquad (8.8)$$

Other compounds may also be formed, depending on the reaction kinetics and on the presence of impurities in the raw materials. The thermal decomposition yields a carbon-rich condensed phase and a hydrogen-rich gas phase. The carbonaceous solid product can either be sequestered or used as material commodities under less severe CO_2 restraints. They can also be used as reducing agents in metallurgical processes. The hydrogen-rich gas mixture can be

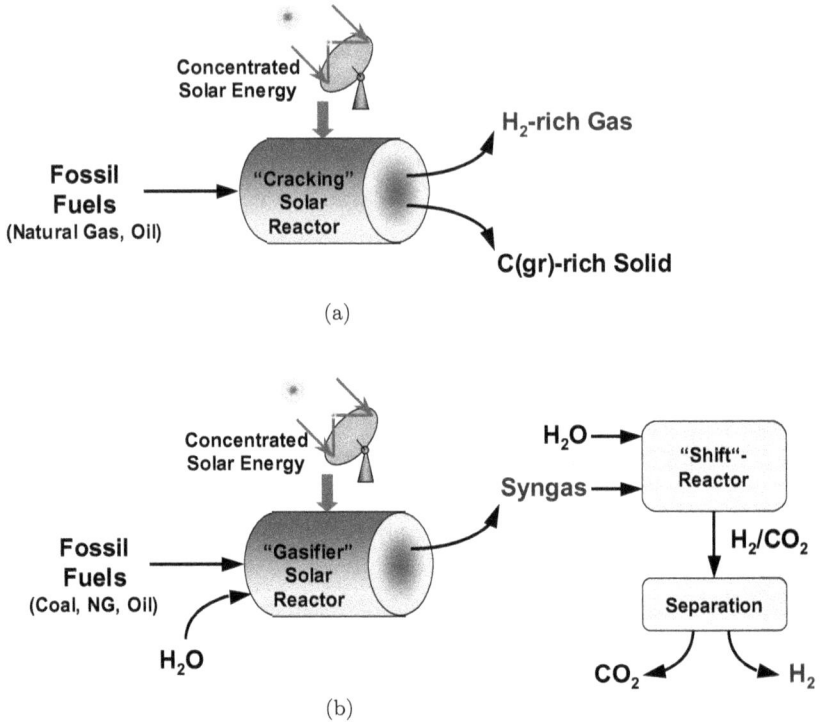

Figure 8.2.5: Simplified process flow diagram for the solar thermal decarbonization of fossil fuels. Two methods are considered: (a) the solar thermal decomposition; (b) the solar thermal steam-reforming/gasification. Omitted is the formation of by-products derived from impurities present in the feedstock

further processed to high-purity hydrogen that is not contaminated with carbon oxides and can be used in fuel cells without inhibiting platinum-made electrodes. H_2-rich mixtures can also be adjusted to yield high-quality syngas.

The steam-reforming/gasification of natural gas, oil, coal, and other hydrocarbons can be represented by the simplified reaction:

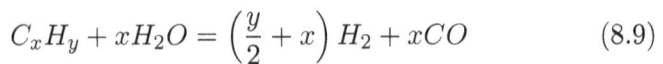

$$C_x H_y + x H_2O = \left(\frac{y}{2} + x\right) H_2 + xCO \qquad (8.9)$$

Other compounds may also be formed, especially with coal, but some impurities contained in the raw materials are cleaned out prior to the decarbonization process. The principal product is high-quality syngas. The CO content can be shifted to H_2 via the catalytic water-gas shift reaction, and the product CO_2 can be separated from H_2 using, for example, the pressure swing absorption technique.

Reactions (8.8) and (8.9) proceed endothermically in the 800–1,500 K range. Some of these processes are currently practiced at an industrial scale, but the energy required for heating the reactants and for the heat of the reaction is supplied by burning a significant portion of the feedstock. Internal combustion results in the contamination of the gaseous products while external combustion results in a lower thermal efficiency because of the irreversibilities associated with indirect heat transfer. Alternatively, using solar energy for process heat offers several advantages: 1) the discharge of pollutants is avoided; 2) the gaseous products are not contaminated; and 3) the calorific value of the fuel is upgraded by adding solar energy in an amount equal to the ΔH of the reaction. Furthermore, by directly irradiating the reactants, solar energy can be efficiently transferred to the reaction site, bypassing the limitations imposed by heat exchangers.

The following routes for H_2 and power generation are examined [29,30]:

1. The solar thermal decomposition of natural gas followed by carbon sequestration and H_2 use in a 65%-efficient H_2/O_2 fuel cell;
2. The steam gasification of coal followed by syngas processing to H_2 (by water-shift gas reaction and H_2/CO_2 separation), which is used to fuel a 65%-efficient fuel cell.

The exergy efficiency for each of these open-cycle routes is defined as the ratio of the work output by the fuel cell to the total thermal energy input by solar and by the heating value of the reactants:

$$\eta_{exergy} = \frac{Work\ Output}{Q_{solar} + HHV_{reactant}} \tag{8.10}$$

where $HHV_{reactant}$ is the high heating value of the fossil fuel being processed, e.g. about 890 $kJ \cdot mole^{-1}$ for natural gas, and 35,700 $kJ \cdot kg^{-1}$ for anthracite coal. Energy efficiencies are calculated for a solar cavity-receiver/reactor operated at 1,350–1,500 K and subjected to a mean solar flux concentration ratio in the range of 1,000–2,000. For route No. 1, aimed at H_2 generation from natural gas, the energy efficiency amounts to 30%. This route offers zero CO_2 emissions as a result of carbon sequestration. However, the energy penalty for completely avoiding CO_2 amounts to 30% of the electrical output, vis-à-vis the direct use of CH_4 for fueling a 55%-efficient combined Brayton–Rankine cycle. Higher energy efficiencies (exceeding 65%) could be obtained when the carbon is either steam-gasified to syngas in a solar gasification process and the syngas further processed to H_2, or used as a reducing agent of ZnO in a solar carbothermal process for producing Zn and CO that are further converted via water-splitting and water-shifting to H_2. Any of these two alternative solar processes yield 2 additional moles of H_2 per mole C(gr) and offer a net gain of 40% in the electrical output (and, consequently, an equal percent reduction in the corresponding specific CO_2 emissions), as compared to the conventional combined cycle power generation. For route No. 2, aimed at H_2 generation from coal, the energy efficiency amounts to 46%. This route offers a net gain in the electrical output by a factor varying in the range 1.7–1.9 (depending on the coal type), vis-à-vis the direct use of coal for fueling a 35% efficient Rankine cycle.

Specific CO_2 emissions amount to 0.49–0.56 kg CO_2/kWh$_e$, about half as much as the specific emissions discharged by conventional coal-fired power plants.

8.2.5 *Conclusions*

Solar thermochemical processes offer an efficient path for storage and transportation of solar energy. The products are renewable fuels for delivering clean and sustainable energy services. There is a pressing need to develop greenhouse gas mitigation options that can be applied to fossil fuels in the mid-term. Solar/fossil fuel hybrid chemical processes for upgrading and converting fossil fuels to hydrogen conserve fossil fuels, reduce emissions, and could become an important transition path towards solar hydrogen. The production of solar hydrogen via 2-step thermochemical cycles based on metal oxides redox reactions have favorable long-term potential, warranting further development and large-scale demonstration.

8.2.6 *Acknowledgments*

This work is supported by the ETH-Project No. TH-29'/99-4 and by the BFE-Swiss Federal Office of Energy Project "Clean Energy Technologies for CO_2 Mitigation".

8.2.7 *References*

[1] Steinfeld A., Palumbo R., Solar Thermochemical Process Technology, *Encyclopedia of Physical Science and Technology*, R. A. Meyers Ed., Academic Press, Vol. 15, pp. 237–256, 2001.

[2] Fletcher E. A, Moen R. L., *Science* 1977; 197: 1050–1056.

[3] Steinfeld A., Schubnell M., *Solar Energy* 1993; 50: 19–25.

[4] Kogan A., *Int J Hydrogen Energy* 1998; 23: 89–98.

[5] Ihara S., *Int J Hydrogen Energy* 1980; 5: 527–34.

[6] Fletcher, E. A., *Ind Eng Chem Res* 1999; 38: 2275–82.

[7] Funk J., *Int J Hydrogen Energy* 2001; 26: 185–90.

[8] Steinfeld A, Kuhn P., Reller A., Palumbo R., Murray J., Tamaura Y., *Int J Hydrogen Energy* 1998; 23: 767–74.

[9] Bilgen E., Ducarroir M., Foex M., Sibieude F., Trombe F., *Int J Hydrogen Energy* 1977; 2: 251-7.

[10] Nakamura T., *Solar Energy* 1977; 19: 467–75.

[11] Palumbo R., Lede J., Boutin O., Elorza Ricart E., Steinfeld A., Möller S., Weidenkaff A., Fletcher E. A., Bielicki J., *Chemical Engineering Science* 1998; 53: 2503–18.

[12] Sibieude F., Ducarroir M., Tofighi A., Ambriz J., *Int J Hydrogen Energy* 1982; 7: 79–88.

[13] Steinfeld A., Sanders S., Palumbo R., *Solar Energy* 1999; 65: 43–53.

[14] Ehrensberger K., Frei A., Kuhn P., Oswald HR., *Solid State Ionics* 1995; 78: 151–60.

[15] Tamaura Y., Steinfeld A., Kuhn P., Ehrensberger K., *Energy* 1995; 20: 325–30.

[16] Moeller S., Palumbo R., *Chemical Enegineering Science* 2001, 56: 4505–4515.

[17] Weidenkaff A., Reller A., Wokaun A., Steinfeld A., *Thermochimica Acta* 2000; 359: 69–75.

[18] Weidenkaff A., Reller R., Wokaun A., Steinfeld A., *Chemistry of Materials* 2000; 12: 2175–81.

[19] Elorza-Ricart E., Martin P., Ferrer M., Lede J., *J Phys IV France* 1999; 9: 325–30.

[20] Steinfeld A., *Int. J. Hydrogen Energy* 2002; 27:611–19.

[21] Haueter P., Moeller S., Palumbo R., Steinfeld A., *Solar Energy* 1999; 67: 161–7.

[22] Berman A., Epstein M., *Int J Hydrogen Energy* 2000; 25: 957–67.

[23] Murray J. P., Steinfeld A., Fletcher E. A., *Energy* 1995; 20: 695–704.

[24] Steinfeld A., Fletcher E. A., *Energy* 1991; 16: 1011–19.

[25] Steinfeld A., Frei A., Kuhn P., Wuillemin D., *Int. J. Hydrogen Energy* 1995; 20: 793–804.

[26] Steinfeld A., Brack M., Meier A., Weidenkaff A., Wuillemin D., *Energy* 1998; 23: 803–14.

[27] Steinfeld A., Kuhn P., Karni J., *Energy* 1993; 18: 239–49.

[28] Kräupl S., Steinfeld A., *J.Solar Energy Engineering* 2001; 123: 237–43.

[29] Hirsch D., Epstein M., Steinfeld A., *Int. J. Hydrogen Energy* 2001; 26: 1023–33.

[30] Zedtwitz V., Steinfeld A., The solar thermal gasification of coal — Energy conversion efficiency and CO_2 mitigation potential, *Energy*, submitted.

8.3   Discussion Following Aldo Steinfeld's Presentation

Roy: What material did you use for your CPC reflectors?

Steinfeld: We used diamond-turned aluminium. But there are other ways of doing it, e.g. with silver-coated steel. The Weizmann Institute has built a very large CPC with glass facets. In fact, for large reflectors one will probably always have to employ facets.

Karni: I would just like to point out that it was Rotem Industries Ltd who performed all of the design analysis and who actually built the Weizmann Institute's CPC.

Roy: What kind of parasitic losses are associated with the use of a secondary reflector?

Steinfeld: Our CPC is water-cooled because no material has 100% reflectivity. So I don't know whether this 5 – 7 % energy loss should be regarded as "parasitic" or simply one of the contributory factors to the overall system efficiency.

Goldstein: Does your use of zinc at high temperatures give you corrosion problems?

Steinfeld: Indeed. Zinc is very corrosive. To begin with, we had severe problems at the contact between molten zinc and steel. However, we solved them by lining the inner surface of our solar reactor with zinc oxide. This way, the hot zinc doesn't have any contact with the steel part of the chamber.

Goldstein: And what about the quartz window. Isn't there a tendency for zinc vapor to condense on its inner surface?

Steinfeld: This is also a problem, and one that is not easy to solve in an energy-efficient manner. On the one hand, we employ aerodynamic protection of the window, by using tangential and radial streams of an inert gas, such as nitrogen and argon. However, both of these gases are fairly energy-intensive in their production. We therefore employ computational fluid dynamics in order to minimize the amounts of these gases that are needed.

Goldstein: Couldn't you employ some of the hydrogen you produce, and then re-cycle it?

Steinfeld: That's an intriguing possibility but it would be complicated due to the presence of oxygen in our system. So, I can't see a simple way of doing this.

Chapter 9

2015

9.1 Editor's Foreword

The large jump in time from 2002 until 2015 merely indicates that the Sede Boqer Symposia from the 12th through the 18th were dominated by photovoltaic and other non-thermal keynote presentations. Thus, in this final chapter of our volume on the solar-thermal aspect of the history of solar power production, we reproduce a reconstruction of the keynote lecture given by Chemi Sugarmen in the 19th symposium, on what in Israel is known as "The Ashalim Project". The name actually refers to 3 parallel solar projects located in the Negev Desert, adjacent to the agricultural village of Ashalim. The projects are important because they constitute three very large (by present-day standards) systems, employing three different solar technologies, all located at the same place, hence exposed to the same ambient conditions, and connected to the electricity grid. Two are solar-thermal (120 MW$_e$ trough-based, and 120 MW$_e$ tower-based, respectively) and the third is a 30 MW photovoltaic project, a further 40 MW being planned. These systems will accordingly provide invaluable information about the moment-by-moment advantages and disadvantages of each under all kinds of weather situation, and the ease with which their parallel but momentarily differing outputs can be integrated with grid requirements.

9.2 "300 MW of Solar Power at Ashalim" (Eng. Chemi Sugarmen)

A keynote lecture presented by Chemi Sugarmen (CTO, Shikun Binui Renewable Energy, Tel Aviv, Israel).

9.2.1 *Disclaimer*

I have been invited to present an overview of the three large-scale solar power projects that are currently under construction adjacent to Ashalim village. Such an overview is complicated by the fact that each of the projects is privately financed by organizations that are in commercial competition with one another. Consequently much of the information we should all like to know is not freely available. A second complication is that I represent one of those organizations. This means that far more details will be given regarding the parabolic trough system than for the others. Regarding the latter, I have to thank the *Megalim* consortium for providing the information I shall present on the solar tower project, and the *Clal Sun* consortium for data on the first stage of the PV project. In spite of the fact that most of the information presented here pertains to the parabolic trough project, much of it is of relevance to at least the other solar-thermal project. So, this emphasis on trough technology should not be interpreted as an attempt to persuade you that our project is more important than the others, even though for us it is!

9.2.2 *Pre-history of the "Ashalim project"*

In 2002 the Israel Ministry of Infrastructure established a professional committee to recommend a suitable location in the Negev Desert for a large-scale solar-thermal plant [More commonly referred to as a Concentrated Solar Power (CSP) plant]. I believe that Professor Faiman was a member of that committee.

In 2004 the government decided upon Ashalim as the location. See Fig. 9.2.1.

The same year, the National Board for Planning and Building was given the task of conducting an "Environmental Impact Assessment" (EIA), and formulating a "Notice of Preparation" (NOP) which was issued as NOP 10/B/1.

In 2007 the government requested the Ministry of Infrastructure to explore the possibility of building a 250 MW solar park at Ashalim.

In 2008 the government decided to issue a "Build, Operate, Transfer" (BOT) type of bid for two solar-thermal plants in the range

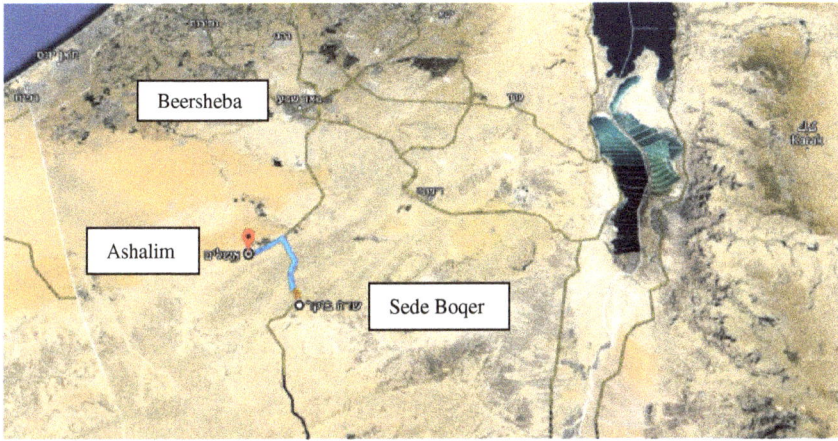

Figure 9.2.1: The location of Ashalim village. Also shown are Sede Boqer (13 km in linear distance) and Beersheba

80–125 MW and two PV plants each of rating 15 MW. The bid was published the following year: in its initial form (but later modified) for CSP and in its final form for PV. Figure 9.2.2 Shows the two CSP plots "A" and "B" and the PV plot, according to NOP 10/B/1, and their locations relative to the agricultural village of Ashalim.

The EIA was submitted in 2011, and in 2013 the government approved the NOP and the EIA.

9.2.3 *The call for proposals*

The terms of the government call for proposals (CFP) required the latter to come from consortiums representing a minimum of one Israeli company and one overseas company with experience in the construction and operation of large-scale solar plants. The CSP proposed projects needed to satisfy the following criteria:

- "Build, Operate, Transfer" (BOT)
- 3 years for construction and then 25 years of operation under a "Power Purchase Agreement" (PPA)
- Maximum annual usage of natural gas = 15%, where the natural gas ratio, NG is defined as:

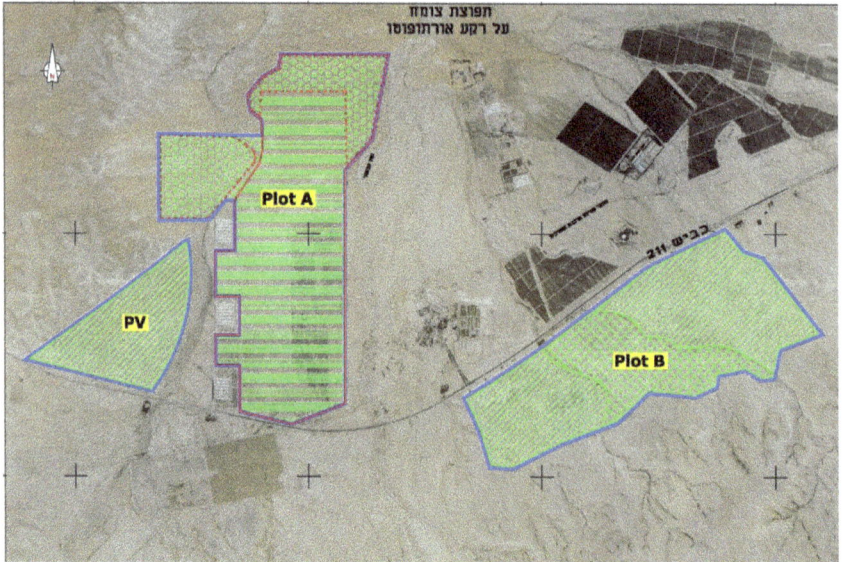

Figure 9.2.2:   Map of CSP plots A and B and plot PV according to NOP 10/B/1. Ashalim village is in the top right corner

NG ratio = (Total gas consumed × LHV) / (Total net electricity × 10,310)

[Here, LHV is the "lower heating value" of the gas, and the factor 10,310 balances the calorific and electrical units so that the ratio is dimensionless].

- Maximum NG ratio per day = 50%
- Maximum NG flow rate set for 50% of maximum power
- Tariff: 0.76 − 0.78 NIS/kWh for power P ≤ 110 MW
  And up to 760 h for 110 < P ≤ 121 MW
- Load plan:
  A minimum of 65% during High and Intermediate Priority hours
  A minimum of 50% during High Priority hours
- Liquidated damages to the State in case of:

  Over NG consumption — annually or daily
  Shortfall in weather-adjusted annual net output

Shortfall in annual High and Intermediate Priority hours
Over annual weather-adjusted parasitics

- Guaranteed net power measured at Israel Electric Corporation (IEC) substation 3 km from the plot
- Zero effluent

In 2011, three contracts were awarded: The *Clal Sun* consortium received a signed concession for the first 30 MW of PV in April 2012; The *Megalim* Consortium received the Plot B concession for 120 MW of CSP (central receiver + heliostats technology) in December 2012; The *Negev Energy* consortium received the plot A concession for 120 MW of CSP (parabolic trough technology) in September 2013.

9.2.3.1 Plot PV1 (single-axis tracking): The clal sun consortium

- Owners: Clal Sun (50%) and Sun Edison (50%), responsible for "Engineering, Procurement, Construction" (EPC) and "Operation and Maintenance" (O&M).
- Lenders: Bank Hapoalim and Deutsche Bank.

9.2.3.2 Plot B (central receiver CSP): The megalim consortioum

- Owners: Brightsource (25.05%) and Alstom (25.05%), responsible also for EPC and O&M, and Noy (an infrastructure and energy investment fund) (49.9%).
- Lenders: Bank Hapoalim and EIB

9.2.3.3 Plot A (parabolic trough CSP): The Negev Energy consortium

- Owners: Shikun Binui (50%) and Abengoa (50%), responsible also for EPC and O&M.
- Lenders: OPIC, EIB and Bank Leumi.

9.2.4   Technical details

9.2.4.1   *Plot PV1*

Plot PV1 is the plot that was assigned to the Clal Sun — Sun Edison consortium for a 30 MW PV system. It occupies approximately one third of the total land area that has been allocated to PV, as seen in Fig. 9.2.3.

At the time of this review, the consortium has not yet finalized the technical details of their system, other than that the PV modules will track the sun about a north-south horizontal axis. Figure 9.2.4 is their simulated picture of how the system will appear.

9.2.4.2   *Plot B*

Plot B was allocated to the Megalim consortium for a 120 MW central receiver CSP system. The principal components of their system are shown, schematically, in Fig. 9.2.5.

The system will consist of a tower, which, including its steam-generating receiver structure, will reach a height of 240 m. The surrounding field will contain 50,600 heliostats.

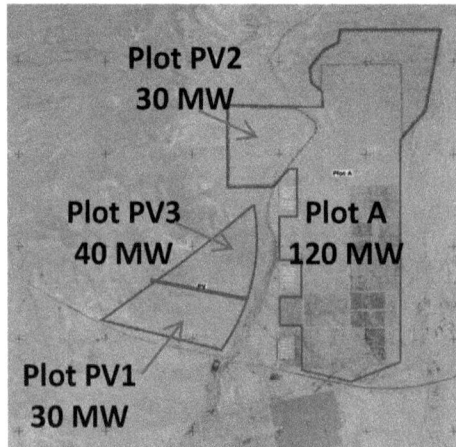

Figure 9.2.3:   Map of CSP plot A and the adjacent plots allocated to PV

Figure 9.2.4: Simulation of the Clal Sun — Sun Edison 30 MW PV system

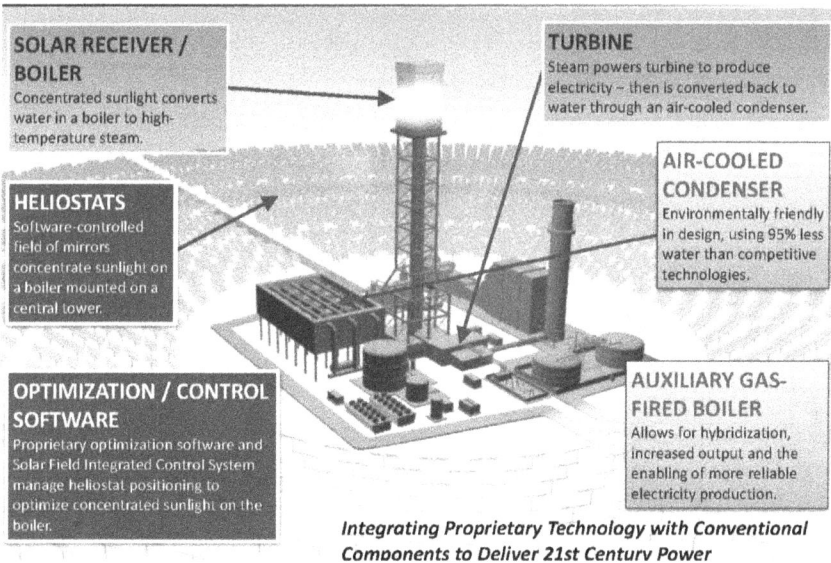

Figure 9.2.5: Principal components of Megalim's central receiver CSP system

Concerned with the aesthetics of the environment, the consortium held an architectural competition for an envelope to surround the tower. It was won by architects Eran Ziv and Izhar Kedmi. A simulation of the tower envelope, the solar steam receiver, and part of the surrounding field is shown in Fig. 9.2.6.

Figure 9.2.6:   Simulated appearance of the Megalim Tower and solar receiver

Figure 9.2.7:   Simulated aerial view of the Megalim SCP plant

9.2.4.3   *Plot A*

Plot A was allocated to the Negev Energy consortium for a 120 MW parabolic trough CSP system. The principal components of their system are shown, schematically, in Fig. 9.2.8.

Table 9.2.1 presents a summary of the Negev Energy CSP plant characteristics.

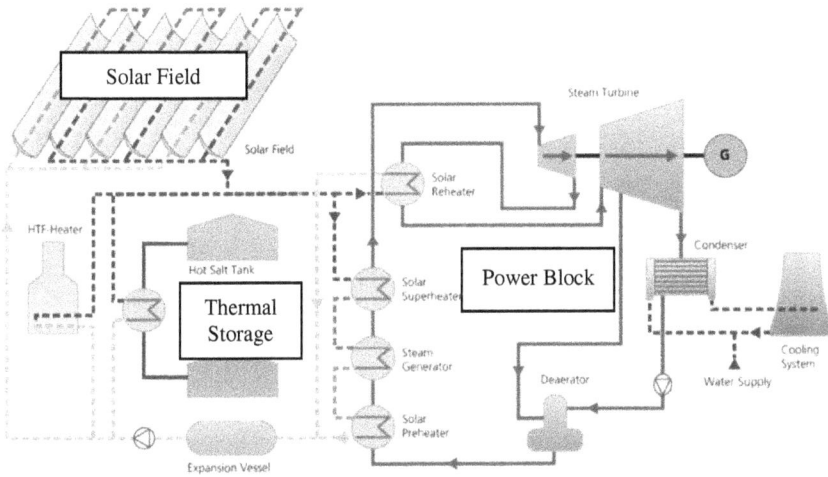

Figure 9.2.8: Principal components of Negev Energy's parabolic trough CSP system

Table 9.2.1: General plant configuration of the Negev Energy CSP plant

Solar field technology	Parabolic trough
Number of loops	338
Solar collectors	4 collectors per loop. Total: 1,352
Number of mirrors	28 mirrors per module. Total: 454,272
Total plant area	~4,000 dunam [400 ha]
Nominal/Maximum electrical capacity	110/121 MW
Cooling system	Wet cooling tower
Thermal energy storage (TES) system	Molten salt, two-tank system
TES capacity	1,440 MWh$_{th}$ (4.5 h full capacity)
Fossil fuel backup	Natural gas, up to 15%

The terms "module", "collector", "loop" employed in Table 9.2.1 are best perceived in the simulated photograph shown in Fig. 9.2.9. The three absorber tubes in the forefront of the picture are illuminated by 28 mirrors. This unit is referred to as a "module". A string of 12 such modules constitute a "collector", the entire structure being driven by a single motor. 4 collectors constitute a so-called "loop".

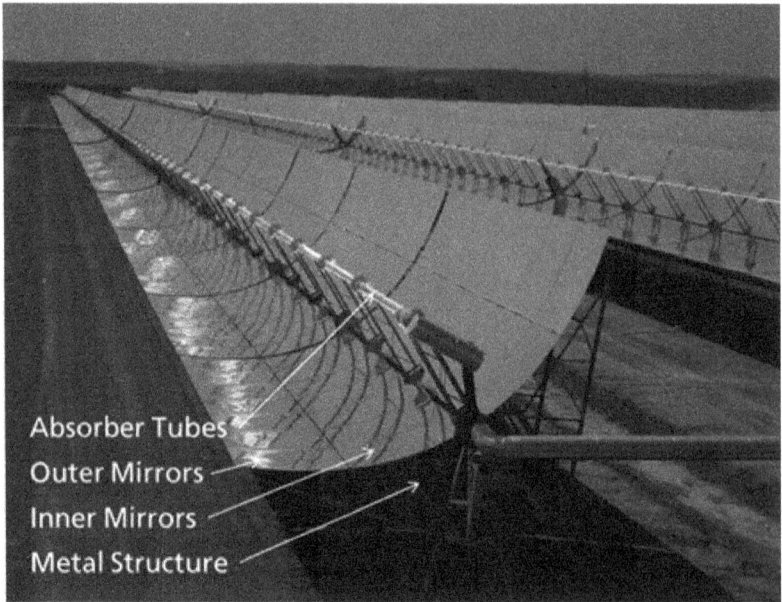

Figure 9.2.9: Simulated photograph showing a number of parallel loops in the Negev Energy CSP system. The front-most "module" is seen to consist of 28 mirrors and three absorber tubes. 12 of these modules constitute a "collector", and 4 collectors complete a "loop"

Figure 9.2.10 shows a schematic layout of the Negev Energy CSP plant. The power block is in the center, surrounded by the loops of parabolic trough solar collectors. The green matrix at the left center contains the plant entrance gate and a temporary logistics area. The three white rectangles below the gate represent evaporation ponds.

Figure 9.2.11 shows a schematic layout of the Power block at the center of the Negev Energy CSP plant.

The principal components of the Thermal Energy Storage (TES) system are two large storage tanks containing hot salt and cold salt, respectively. Electrical immersion heaters ensure that the salt in both tanks is liquid at all times. Heat exchangers enable transfer of surplus thermal energy from the collectors' heat transfer fluid (HTF) to the salt in the cold tank, and transfer of thermal energy from the salt in the hot tank to the steam turbine generator at times when direct

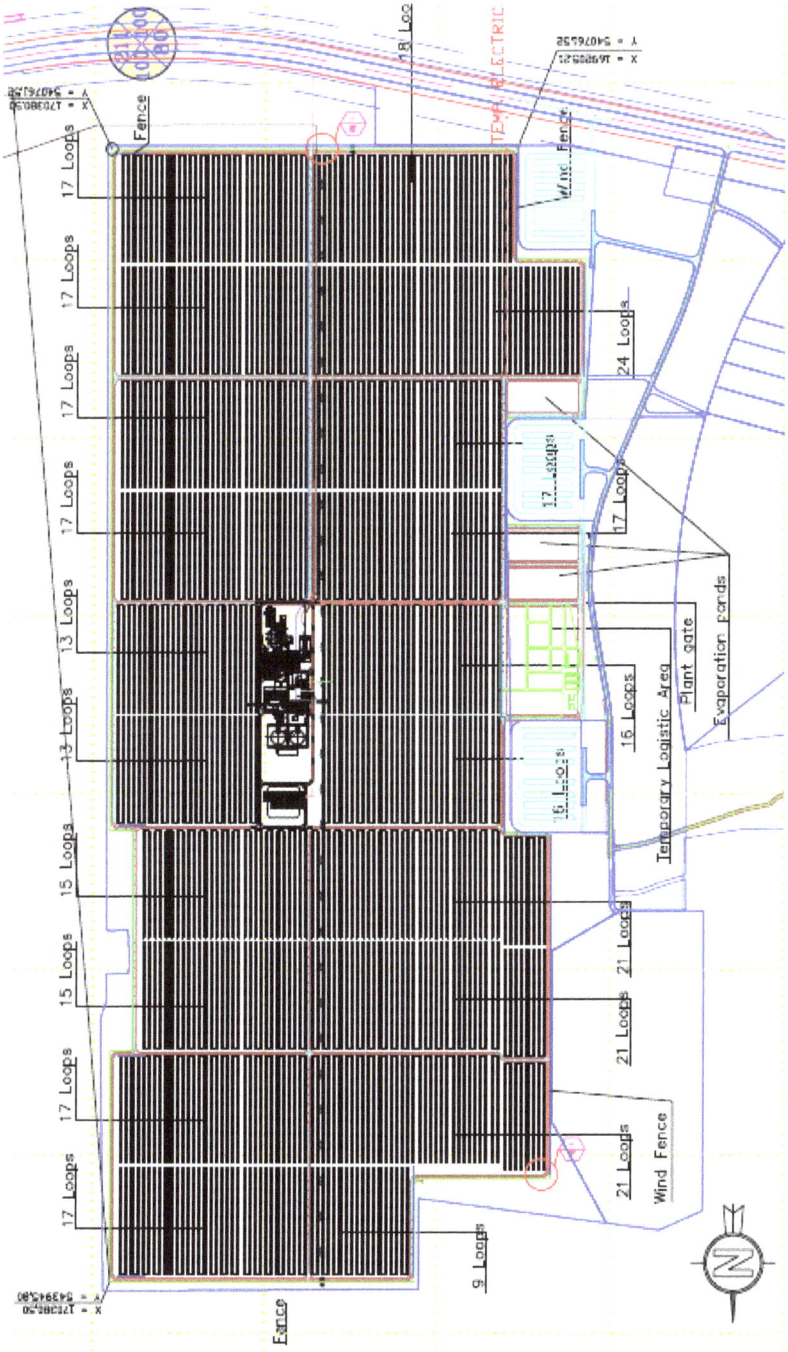

Figure 9.2.10: Negev Energy CSP plant schematic layout

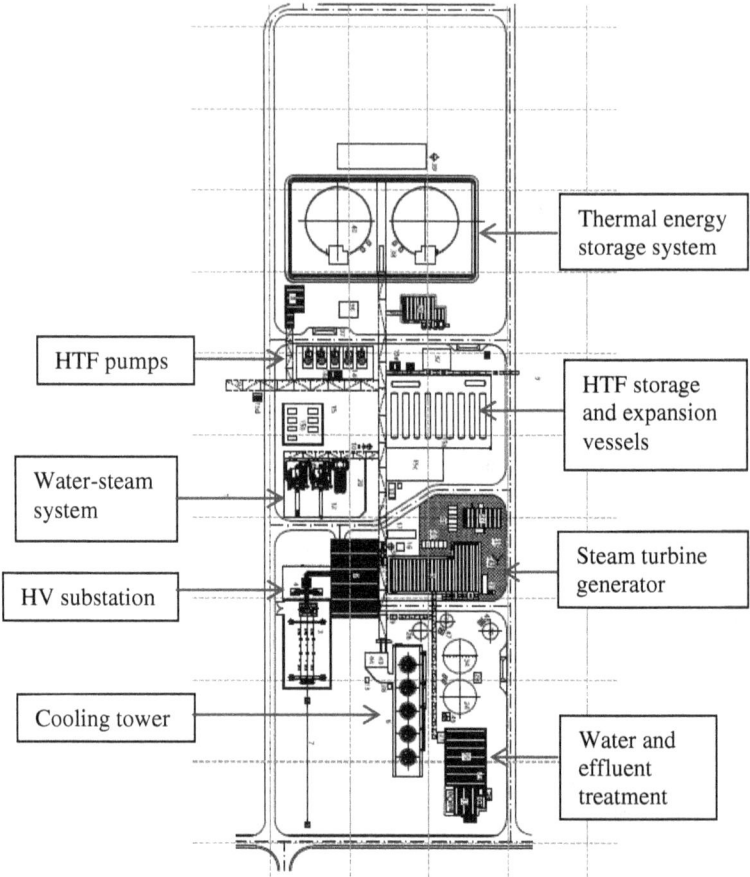

Figure 9.2.11:   Negev Energy power block schematic layout

input from the solar collectors is inadequate. Pumps transfer the heated salt from the cold tank to the hot tank, and return the energy-depleted salt from the latter to the cold tank.

Thermal energy storage expands the generation of solar electricity beyond the period when the sun is actually shining.

- It allows the time-shifting of generation to respond to changes in supply and demand in similar manner to the function of conventional thermal generation.

- It allows the CSP plant to store excess energy collected by the solar field, and discharge it later when the solar resource is low. This effectively increases the solar fraction of the plant, as this excess energy is not wasted.
- It enables electricity generation during critical peak hours with high loads. This has a significant impact on capacity credit, and demonstrates the eventual ability of renewable energy generation to replace fossil fuel. At the present stage of the technology, storage capacity is limited to 4.5 hours, but, with further developments, there would be additional benefits for 6–9 hours of storage.

Figure 9.2.12 shows an artist's impression of an aerial view of the finished plant, looking north-eastward from the south-west corner.

9.2.5 *Technology comparison*

Each of the solar power technologies that will be demonstrated at Ashalim has a number of known advantages and disadvantages compared to the others, and in all probability, also some presently unknown ones that will become apparent when all systems are operative.

Figure 9.2.12:   Simulated aerial view of the Negev Energy CSP plant

9.2.5.1   *CSP (+ TES) vs. PV*

The advantage of PV compared to CSP in regions where skies are mostly overcast is well-known and needs no elaboration here. Also well established, is that technology's enormous modularity (home electronics to power plants) and the greater simplicity of its systems compared to CSP. However, less obvious are the several advantages that CSP technologies have over PV for large-scale power plants in desert regions such as Israel. Specifically:

- CSP is a fully "dispatchable" renewable energy technology, allowing the provision of electricity at times when it is valued the most highly, namely during peak or high priority hours.
- Dispatchability is particularly important for Israel because: it is an "island" of generation; with a terrain that offers no natural storage facilities; and the absence of carbon-free base-load generation (such as hydroelectric or nuclear power plants).
- Dispatchability gives CSP a higher capacity value than PV, with a higher premium for availability than the latter.
- Dispatchability of CSP results in reduced need of backup from conventional power plants.
- "Thermal momentum" (i.e. heat collected and absorbed into the HTF) allows the electrical output of the plant to remain constant (or change very gradually) during periods of partial cloud cover, or at sunset, allowing utilities time to shift to alternative sources of generation.
- CSP plants are easier to hybridize with conventional power projects and fuels.
- PV imposes costs on the grid (caused by its uncertain, variable and unpredictable production).
- CSP presents greater employment opportunities, both during the construction and operation phases.
- The related R&D activities allow higher levels of local involvement and foreign direct investment.

9.2.5.2 *Parabolic trough vs. central receiver*

It has long been acknowledged that Central receiver technology holds out much promise for the future of CSP power production owing to the potential of such 3-D concentrators to reach much higher temperatures than 2-D Parabolic troughs can achieve. However, many materials issues will need to be solved before Tower-based systems can operate reliably in the temperature range (around 1,000°C) where full advantage can ensue from the resulting higher efficiencies. So, for the time being, there are still a number of advantages that Parabolic troughs have over Central receivers. These are:

- Parabolic trough systems have been in wide use for utility-grade power generation since the mid 1980s.
- It is the most mature and commercially proven of the CSP technologies, having dominated the concentrated solar thermal power industry for the past two decades (∼95% of total CSP capacity in operation), and winning the confidence of utilities and investors.
- It has a solid operational and commercial track record of reliable performance, the O&M costs being predictable and well understood. By contrast, only a handful of tower plants are currently operational.
- Parabolic trough systems have the highest peak efficiency.
- The technology has an inherent modularity and scalability — by adding more loops in parallel, the capacity of a plant can be increased.
- "Off-the-shelf" systems are available.
- Trough technology is still relatively more "bankable" than tower technology.

9.2.6 *Summary*

Having now discussed all of the information that is presently available for the three solar power plants scheduled for the Ashalim

Table 9.2.2: Principal features of the three solar plants in the Ashalim Project

	Plot PV	Plot A	Plot B
Technology	PV single axis tracking	Parabolic trough, Rankine cycle,	Heliostats + tower Rankine cycle
Storage	N/A	1440 MW$_{th}$ 4.5 h at 139 MW$_e$,	N/A
Cooling medium	N/A	Water	Air
Land area [dunam]	750	3980	3150
Installed gross power [MW]	33 (estimate)	139	130 (estimate)
Maximum net power [MW]	30	121	121
Annual generation [GWh]	62 (estimate)	Confidential	?
Tariff [NIS/kWh]	0.53	0.76	0.79
Expected service life [y]	20	40	?

Project (and presented some details here for the first time), let me summarize what we have learned. Table 9.2.2 lists in compact form, of all technical and numerical information that is presently public knowledge regarding the three systems.

9.2.7 *Future trends and forecast*

I think that there are several reasons to be optimistic about the future of CSP technologies for the future large scale generation of electricity. In particular, studies of the cost curve indicate a potential cost reduction in the range 25%–55% by the years 2025–2030, owing to:

- *Volume production*: Installed capacity has achieved a compound annual growth rate of over 40% during the past 5 years. A significant uptake is also expected during the coming 20 years.
- *Plant scale-up*: Large capacity CSP plants will offer lower per MW costs and higher efficiencies.
- *Technological advance*: R&D initiatives are currently underway with the objective of increasing operating temperatures and

pressures, to increase efficiency and reduce electricity costs. Such efforts are going into thermal energy storage, super-heated towers, molten salt towers, greater hybridization, and new heat transfer fluids.

- As a result of these cost reductions, the level of tariff support that is needed to make CSP economically viable is expected to decrease.

Afterword

Following on from a brief review of the approximately twenty years of solar-thermal research and development that had gone before, this book has presented, a chronicle of how solar-thermal power production developed during the three decades starting in 1986, when the first 10 MW, solar-trough plant was constructed by the *Luz* corporation at Daggett, California, through 2015, by which time such plants at the scale of 100 MW had become widespread in several parts of the world. The 30-year chronicle is in the words of those scientists and engineers who were largely responsible for having made it happen.

Another solar-thermal technology that was also under consideration back in 1986, namely the solar tower, has also matured, to the extent that one of the 120 MW solar plants at Ashalim village in Israel, which was discussed in our final chapter, was designed by the same *Brightsource* company that currently holds the record for having constructed the world's largest — 392 MW — solar-thermal plant, at Ivanpah in the USA. In passing it is worth mentioning that both *Luz* and *Brightsource* were founded by the late Arnold Goldman (1943–2017), and that both of the Ashalim plants are now on-line.

Because photovoltaic (PV) plants were only mentioned *en passent* in this volume. It is never the less important to note that during this same period of time, PV technology has matured from systems on the scale of 1 MW, of doubtful reliability, to ones in the 1,000 MW range, in China and India, with expected lifetimes of 25 years and more.

Does this mean that PV plants are more cost-effective than CSP (i.e. solar-thermal) plants? Superficially, it might appear so. For on the one hand, PV plants can produce electric power wherever there is daylight, no matter how cloudy or rainy the location may be. This situation alone wins out over CSP systems, which can only operate cost-effectively in sunny desert regions. Furthermore, PV systems come in all shapes and sizes, whereas CSP plants tend to be more cost-effective the larger they are. So, it makes obvious commercial sense that there are far more manufacturers of components for PV systems than there are for the solar-thermal variety. This fact has facilitated a dramatic down-swing in PV module prices in recent years, and led to the current lower dollar-per-watt cost of PV plants compared to the solar-thermal varieties.

However, as was emphasized by several of the specialists in this volume: dollars-per-watt cannot be compared in meaningful manner when different power-generating technologies are at issue. What counts is dollars-per-kilowatt-hour *at times when the electricity is needed*. So, until battery storage of electricity becomes a cost-effective option, solar-thermal technologies will maintain an advantage because the thermal energy they produce as a first stage can be readily stored until needed for electricity generation.

In this regard, there is a second reason why it is too early to dismiss solar-thermal power on a simple dollar vs. dollar basis. In order to grasp this point it is useful to think about almost any contemporary power company, whose various generators use coal, gas and oil (even without renewable energy sources). In today's competitive economies, power companies must generate electricity 24 hours a day, in as economical a manner as possible. To illustrate this fact, Table 10.1, reproduces data published by the Israel Electric Corporation for the year 2015 [1]. The table shows the amounts of the various fuels the company used during that year, and the cost of the electricity produced by each fuel type.

A quick glance at the table would suggest that on economic grounds the company should have used 100% coal for a cost of

Table 10.1: Fuel usage by the Israel Electric Corporation during the year 2015

Fuel type	% Usage	Electricity cost [Ag/kWh]
Coal	57.6	12.36
Natural gas	40.3	18.74
Liquid gas	1.3	71.81
Diesel oil	0.7	110.23
Fuel oil	0.1	147.18

12.36 Ag/kWh, and not wasted money on generation with any of the other fuels. However, had such a strategy been adopted, the company would have needed to generate considerable quantities of electricity at night, and at other times of low demand. This is because the output of coal-powered generators cannot be rapidly adjusted to suit the moment-to-moment needs of customers. The strategy of the company was therefore to use all 5 types of generators in its fleet, but in amounts that enabled it to provide for all needs at a maximum profit for the investors. Elementary manipulation of the columns in Table 10.1 shows that the resulting average cost of electricity from such a strategy turned out to be 16.52 Ag/kWh [=4.25 US¢/kWh].

The point in the above illustration is that no single generator type is capable of producing 24 hour per day electricity in a cost-effective manner. One must always employ a mix of technologies, each having its own characteristic response time to changes in demand, but equally relevant, each having its own characteristic cost. This is the second reason why it is premature to dismiss CSP, in either of its current technological forms, in favor of apparently lower-cost PV. To this end, an extremely important feature of the Ashalim project, discussed in chapter 9, is that it will enable future decision makers to understand the response times of all three technologies, when operated under identical climatic conditions, *together* with their respective costs. It is to be hoped that some appropriate mix, perhaps together with wind generators and some yet-to-be-developed storage systems, will enable the complete elimination of fossil fuel from a future electrical distribution system.

Finally, it should not be forgotten that a third technically excellent solar-thermal technology — the salt-gradient solar pond, with its built-in multi-hour storage properties — discussed in chapter 2, was dismissed on economic grounds that were deemed important at the time. It should not be ruled out that such a technology may yet have a role to play in that future much desired fossil-free electricity grid.

Reference for Afterword

[1] *Israel Electric Corporation Ltd. Financial Reports For The Year Ended December 31, 2016.* https://www.iec.co.il/EN/Documents/The_Israel_Electric_Co-Financial_Reports_December_2016b.pdf (visited 12.6.2018).